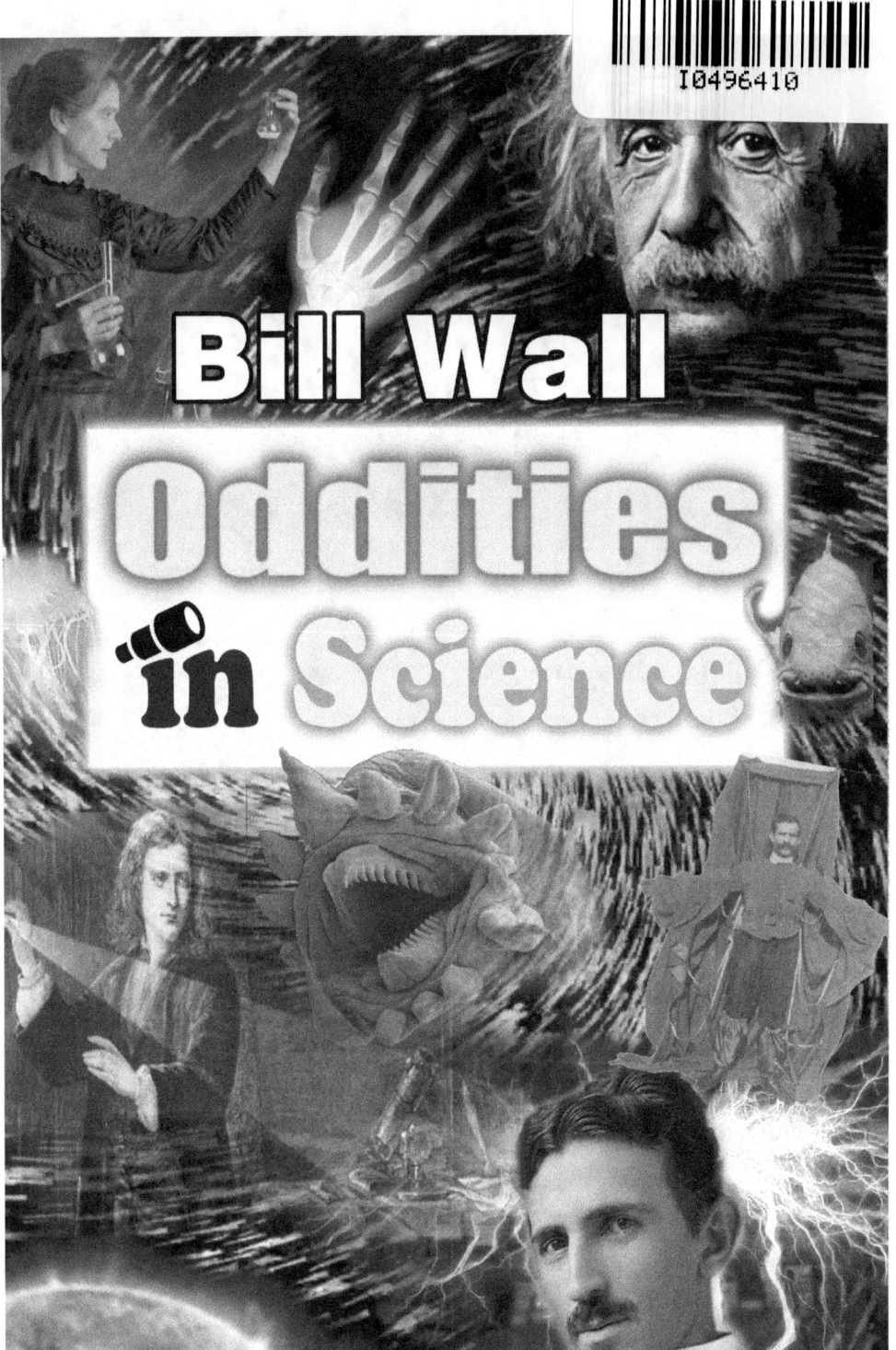

© Copyright 2020
All rights reserved

Cover Design/Layout
Gerald Lee Wall

ISBN: 9798636066040
http://www.billwallchess.com/

Table of Contents

Oddities in Astronomy 1
Oddities in Biology 9
Oddities in Chemistry 17
Oddities in Computer Science 25
Oddities in Geology 33
Oddities in Medical Science 41
Oddities in Meteorology 49
Oddities in Physics 55
Oddities in Zoology 61
Oddities of Discovery 69
    Accidental Discoveries 69
    Odd Discoveries 75
Oddities of Scientific Studies 79
Oddities of Scientific Terms 87
Oddities of Scandals and Frauds 91
Oddities in Science Fiction 101
Oddities in Myths 105
Oddities of Scientists 111
    Eccentric Scientists 111
    Murdered Scientists 114
    Death of Scientists 120
    Suicides of Scientists 129
    Scientists Who Disappeared 139
Oddities of the Nobel Prize 145
    Nobel-Prize Winning Scientists 145
    Scientists Who Should Have Won a Nobel Prize 158
Oddities of Criminal Scientists 167

Introduction

This is a collection of oddities about science and scientists. It is the odd, the weird, the strange, the bizarre, the curious, the eccentric, the unusual, the tragedies, the trivia, and the humor in science.

The main scientific fields where oddities are found include astronomy, biology, chemistry, computer science, geology, medical science, meteorology, physics, and zoology. There are also oddities from accidental scientific discoveries or mysterious discoveries that can't be explained.

Included are some of the oddest scientific studies that have ever been conducted. Some of the studies include those involving cadaver arms for punching, calories in cannibalism, clams on Prozac, elephants on LSD, frog odors, knuckle cracking, sword swallowing, and the dynamics of buttered toast tumbling from a table to the floor.

A list of odd scientific terms and measurements is included to keep you abreast in a variety of scientific disciplines that measure in barleycorns, barns, bohrs, cubits, FLOPS, googols, mickeys, moles, quads, racks, rads, and shakes.

There is a section on scandals and frauds in science, with plenty of examples from unscrupulous scientists and pseudo-science projects and programs.

Many of today's scientific discoveries, such as antidepressants, the atomic bomb, cell phones, communications satellites, credit cards, man on the moon, robots, smartwatches, etc., were predicted in the early stories from science fiction.

Science myths are also included and explained. Bats aren't blind. You use more than 10% of your brain. Bulls don't get angry at seeing red; they are color-blind. Carrots will not help with night vision. You don't need to wait 30 minutes or an hour after eating before you go swimming. And elephants are not scared of mice.

Eccentric scientists are not ignored. There are plenty of them. Also included are murdered scientists, accidental deaths, suicides, and the disappearance of scientists. And then there are the oddities of Nobel-prize winning scientists and the stories behind the scientists who should have won a Nobel Prize, but didn't for various reasons.

So stay home, relax, read the stories, enjoy the graphics, and be part of the odd people who love science, even with all its oddities and eccentricities.

Oddities in Astronomy

Corona. The corona, the sun's outer atmosphere, is hotter than the sun's surface. It can be heated up to 10.8 million degrees Fahrenheit, whereas the sun's surface is only 10,000 degrees Fahrenheit. For over a century, solar physicists have been mystified by the sun's ability to reheat the corona. The mechanics of coronal heating are currently unknown. It may be related to processes in the sun's magnetic field. The lowest layer of the sun's atmosphere is called the photosphere, which is about 300 miles thick. The sun's equatorial regions rotated in about 24 days, while the polar regions are slower and take more than 30 days to make a complete rotation. [source: Sharp, "Atmosphere of the Sun: Photosphere, Chromosphere & Corona," *space.com*, Nov 2, 2017]

Dactyl. Dactyl is a moon less than a mile in diameter (4,600 feet) and is the natural satellite or moon of asteroid Ida (243 Ida). Dactyl was discovered by mission member Ann Harch in images returned from the Galileo spacecraft. Before this, scientists had no evidence that asteroids could have moons. Since this discovery, 24 more moons have been found to orbit asteroids. The moon is named after the Dactyls, creatures which inhabited Mount Ida in Greek mythology. [source: "Ida and Dactyl," *solarviews.com*]

Eta Carinae. Eta Carinae is a stellar system containing at least 2 stars. Their combined luminosity (radiated power) is more than 5 million times that of the Sun. The primary star was 250 times the mass of the Sun, but it has lost at least 30 solar masses recently and is expected to explode as a supernova. It is the only star known to produce ultraviolet laser emission. The second star is about 80 times as massive as the Sun. 130 years ago, the primary star exploded, but kept the star intact. The original explosion created a huge 10-solar-mass cloud that expanded at more than 20 million miles per hour – fast enough to travel from Earth to Pluto in a few days. [source: "Astronomers stunned again by Eta Carinae – the star that will not die," *Astronomy Now*, Aug 3, 2018]

Equinox. The term equinox is derived from two Latin words – aequus (equal) and nox (night). An equinox is commonly regarded as the instant of time when the plane of Earth's equator passes through the center of the Sun. The equinox is the time of balance, with theoretically 12 hours of sunshine and 12 hours of non-sun. In practice, it is not exactly equal. There is actually more day than night on the day of an equinox. That's because Earth's atmosphere bends (refracts) sunlight upward. Also, the Sun is not a single point of light, but a large disk. Together, these factors add more daylight to the equinox. The real date of sunlight equality is 3 to 4 days ahead of the equinox. [source: "What is an equinox, and why does it happen?" *National Geographic*, March 19, 2020]

Exoplanet. An exoplanet or extrasolar planet is a planet outside the Solar System. As of April 2020, there have been 4,245 confirmed exoplanets in 3,140 star systems, with 691 systems having more than one planet. The methods used to detect an exoplanet include transit photometry, radio velocity method, Doppler spectroscopy, and gravitational microlensing. The least massive planet, named Draugr, is about twice the mass of the moon. The most massive planet, named HR 2562, is about 30 times the mass of Jupiter. In 1988, the first suspected scientific detection of an exoplanet occurred. In 1992, the first confirmation of detection occurred, with an exoplanet revolving around a pulsar. In 1995, the first confirmation of an exoplanet orbiting a main-sequence star occurred. This was detected by Michel Mayor and Didier Queloz of the University of Geneva. The two shared the 2019 Nobel Prize in Physics with James Peebles who contributed theoretical discoveries in physical cosmology. In 2019, an earth-sized planet was found in the habitable zone of a nearby star. It was discovered from data from the Transiting Exoplanet Survey Satellite (TESS). [source: Kopparapu,

Oddities of Science

"Earth-Sized Planet Found in the Habitable Zone of a Nearby Star," *Live Science*, Jan 7, 2020]

Fast Radio Burst. In radio astronomy, a fast radio burst (FRB) is a transient radio pulse that lasts about a fraction of a millisecond. It is one of the oddest puzzles of modern astronomy. It was discovered in 2007 by Duncan Lorimer and his student David Narkevic. FRBs are rare, extremely bright flashes of light with radio wavelengths. Their origin seems outside our Milky Way Galaxy. They seem to come from regions with strong magnetic fields. It is caused by some high-energy astrophysical process not yet understood. In 2020, for the first time, an FRB was found to be repeating, in a regular 16.35-day cycle. The discovery was made by Dongzi Li, an astrophysicist at the University of Toronto, using the Canadian Hydrogen Intensity Mapping Experiment (CHIME) radio telescope in British Columbia. [source: "Periodic activity from a fast radio burst source," *arXiv.org*, Jan 28, 2020]

Gamma-ray Burst. Gamma-ray bursts (GRBs) are the strongest and brightest electromagnetic events (explosions) known to occur in the universe. They are extremely energetic explosions that can last from 10 milliseconds to several hours. GRBs are thought to be released during a supernova (exploding star) and the formation of a black hole. The release of energy in a few seconds of an exploding supernova is roughly equal to that produced by the sun in its entire 10-billion-year lifetime. The short burst GRBs appear to originate from the merger of binary neutron stars. The first GRB was accidently discovered in 1967, when an Air Force satellite called Vela spotted one. Vela was designed to watch for nuclear explosions on earth but ended up finding gamma-rays coming from beyond the solar system (about 3 billion light-years away). [source: Mann, "What Is a Gamma-Ray Burst?" *space.com*, Jan 15, 2020]

Gran Telescopio Cararias. Gran Telescopio Cararias (GTC) is the world's largest telescope that measures over 34 feet (410 inches) across. It is located on the island of La Palma in the Canary Islands of Spain. It took over 1,000 people from over 100 countries and 22 years to complete. Scientific observations began in July

2009. [source: Moreno, "Huge telescope opens in Spain's Canary Islands," phys.org, July 24, 2009]

Haumea. The dwarf planet Haumea (named after the Hawaiian goddess of childbirth) is an odd object. First discovered in 2004, it orbits in the Kuiper Belt, which is beyond Neptune. It has a strange elongated shape with two moons orbiting it. A day on this dwarf planet lasts 4 hours, making it the fastest-spinning large object in our solar system. In 2017, it was discovered that it had extremely thin rings orbiting it, likely due to some collision in the distant past. The dwarf planet is bright as snow. Haumea is the 3rd largest dwarf planet, behind Pluto and Eris. [source: Beatty, "Surprise! Dwarf Planet Haumea Has a Ring," *Sky & Telescope*, Oct 13, 2017]

HD 139139. HD 139139 (also known as EPIC 249706694) is a sun-like star about 350 light-years away. This odd star was observed by the Kepler space telescope to have 28 dips in its light, each of which lasted between 45 minutes and 7.5 hours. The random dips do not appear to be periodic as would be expected if they were due to transiting planets. The first explanation was that this star had as many as 28 planets revolving around it. It also may be due to dust-emitting asteroids or short-lived star spots. It is likely that this is a bound pair of stars, a little larger and brighter than our sun. [source, Starr, "Astronomers Found a Star That Dims So Erratically, They Have No Explanation For It," *Science Alert*, Jul 2, 2019]

Hyakutake. Hyakutake was discovered on January 1, 1996 by the amateur astronomer Yuji Hyakutake (1950-2002) in southern Japan, using 150 mm (6 inches) binoculars. Its passage near the Earth was one of the closest cometary approaches in the past 200 years. Scientists made the first discovery of X-ray emissions coming from a comet. This was believed to have been caused by ionized solar wind particles interacting with the atoms in the coma of the comet. On May 1, 1996, the spacecraft probe *Ulysses* unexpectedly crossed the ion trail of this comet. It revealed that the tail was at least 3.8 AU (over 353 million miles) in length. This is the longest tail known for a comet. The last time that comet Hyakutake was in the inner Solar system was 17,000 years ago. Its next appearance in the inner Solar System will be 70,000 years from now. [source: Howell, "Hyakutake: Comet with a Long, Long Tail," *space.com*, Feb 28, 2013]

Hypervelocity stars. Hypervelocity stars are stars with velocities that deviate substantially from the normal velocity of stars in a galaxy. Such stars may have velocities so great that they exceed the escape velocity of the galaxy. There have been about 20 observed hypervelocity stars. One star, S5-HVS1, is traveling at close to 4 million miles per hour, 10 times faster than any other star detected so far. The star is about 29,000 light-years from Earth, and may have been ejected out of the Milky Way Galaxy after interacting with the Sagittarius A* super massive black hole at the center of the galaxy. The star interacted with the black hole about 4.8 million years ago. [sources: James, "Star Ejected by Milky Way's Supermassive Black Hole Now traveling at over 3 Million MPH," *Outer Places*, Nov 13, 2019 and "Discovery of a nearby 1700 km/sec star ejected from the Milky Way by Sgr A*," *Monthly Notices of the Royal Astronomical Society*, Nov 4, 2019]

Mercury. Mercury is the smallest and most innermost planet. It is the least understood of all planets. It has the largest eccentricity of any planet. Two moons of Saturn are larger than Mercury. Its temperature on the surface can be as cold as 280 degrees below zero Fahrenheit, or as hot as 800 degrees Fahrenheit (Venus is hotter at 863 degrees Fahrenheit). The names for the features of Mercury are odd. Craters are named for dead artists, musicians, painters, and authors. Ridges are named for scientists who have contributed to the study of Mercury. Depressions are named for works of architectures. Escarpments (steep slope or long cliff) are named for ships or scientific expeditions. Valleys are named for abandoned cities, town, or settlements of antiquity. On average, Mercury is closer to Earth than Venus or Mars. Because Mercury is so close to the Sun, the Hubble Space Telescope cannot observe the planet without burning out the optics. Galileo was the first person to observe Mercury with a telescope. A trip to Mercury requires more rocket fuel than is required to escape the Solar System. In 2018, a spacecraft was launched to orbit Mercury. The spacecraft will not reach Mercury until 2025. [source: Jaggard, "Mercury Explained," *National Geographic*, Jan 21, 2020]

Monster Galaxy. Astronomers recently found an odd ultra-massive monster galaxy that existed about 12 billion years ago. Dubbed XMM-2599, the galaxy formed stars at a high rate and then died. Why it suddenly stopped forming stars is unclear. Even before the Universe was 2 billion years old, XMM-2599 formed a mass of more than 300 billion stars. That means it was forming stars at the rate of 1,000 solar masses every year for around 500 million years. By comparison, our Milky Way Galaxy creates 4 or 5 solar masses per year. XMM-3599 then became inactive in less than a billion years. For a long time, astronomers thought that giant galaxies couldn't form in the early Universe. Numerical computer models can now account for massive galaxies like XMM-2599. The predicted galaxies, however, are expected to be actively forming stars. What makes XMM-2599 so odd is that it stopped forming stars all of a sudden. It may be that it stopped getting fuel (hydrogen gas) or a black hole stopped it from forming new stars. [source: Starr, "Astronomers Find Ultramassive Galaxy From The Early Universe That Suddenly Died, *"Science Alert*, Feb 7, 2020]

Pallas. Pallas in an odd asteroid in the asteroid belt, between Mars and Jupiter. It was the second asteroid to have been discovered (after Ceres) and is the 3rd largest asteroid (318 miles in diameter), behind Ceres and Vesta. It was discovered in 1802 by Heinrich Olbers, and first thought to have been a planet. Pallas is named after Pallas Athena, an alternate name for the goddess Athena. Pallas is the most cratered 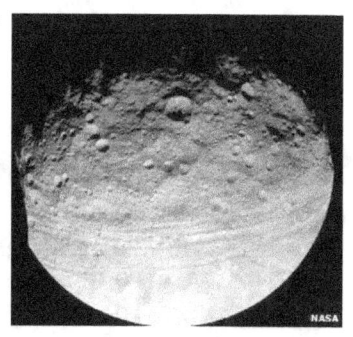 object that we know of in the asteroid belt. It is so pot marked that it looks like a golf ball. [source: "The violent collisional history of aqueously evolved Pallas," *Nature Astronomy*, Feb 10, 2020]

Tabby's Star. Tabby's Star, also known as Boyajian's star, is one of the oddest stars in our Milky Way Galaxy. It has unusual light fluctuations for a star, including up to a 22% dimming in brightness. The discovery was made in 2015 by citizen scientists as part of the Planet Hunter project, using data collected by the Kepler space telescope. No hypothesis to date can fully explain all aspects of the light luminosity curve of this star. The names of the star refer to American astronomer Tabetha Boyajian (1980-) who was the lead author of the paper which investigated the highly unusual light curve. She called it the WTF Star ("Where's the Flux?") after the title of her paper. The star's changes in brightness are consistent with many small masses orbiting the star in tight formation. The first major dip in 2011 showed a reduction of the star's brightness by 15%. In 2013, the reduction in brightness was 22%. By contrast, a planet the size of Jupiter would only reduce the brightness of the star by 1%. One hypothesis that was quickly ruled out was that this was the work of alien mega structures. The latest theory is that the dimming may have been produced by fragments resulting from the disruption of an orphaned exomoon. [source: Young, "What's Going On with Tabby's Star? It's Complicated," *Sky & Telescope*, June 6, 2018]

Tunguska event. The Tunguska event was a large explosion in Siberia in 1908, thought to have been caused by an air burst of a meteoroid or asteroid or comet. The explosion flattened 80 million trees over 830 square miles. Reports indicated that 3 people died in the event. It is classified as an impact event, even though no impact crater has ever been found. A scientific expedition into the area did not occur until 1921. Another expedition took place in 1927. Locals were reluctant to help scientists in the expeditions, believing that the blast was a visitation by their god, who had cursed the area by smashing trees and animals. The Tunguska event is the largest impact event on Earth in recorded history. [source: "Tunguska explosion," *EarthSky*, June 30, 2019]

Uranus. Uranus is an odd planet. It is the only one in our Solar System that is tilted on its side for reasons scientists have not figured out. Its north and south poles lie where most planets have their equators. Its rotation axis is oriented 98 degrees relative to its orbit, whirling around in a clockwise direction. Was it hit by another planet that caused a titanic collision? A new hypothesis was that Uranus had a ring system large enough to cause it to wobble on its axis like a spinning top. In April 2017, astronomers found hydrogen sulfide in its clouds. The planet smells like rotten eggs. Uranus was

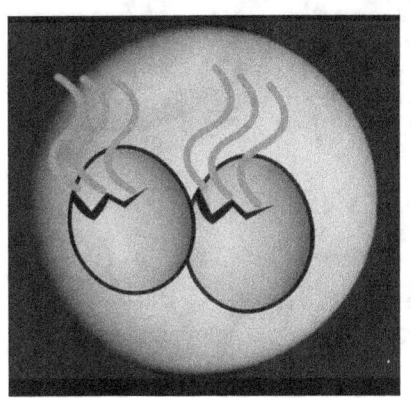

the first planet found with the aid of a telescope in 1781. It was originally thought to be a comet or star. It took two years to decide that it was a new planet. This discoverer, William Herschel, wanted to name it Georgium Sidus, after King George III. Astronomer Johann Bode suggested that it be name after the Greek god of the sky, Uranus. Uranus has the coldest planetary atmosphere at -371 Fahrenheit (-224 Celsius or 49 Kelvin). The wind speeds of the clouds can reach up to 560 miles per hour. [source: Starr, "There's a New Hypothesis For How Uranus Ended Up Tipped on its Side," *Science Alert*, Mar 13, 2020]

UY Scuti. The red hyper giant pulsing variable star UY Scuti is the largest star known. It is located in the constellation Scutum (the shield) in the southern sky. It has a radius about 1,700 times larger than the Sun (650 million miles in diameter) and about 10 times heavier. It is 340,000 times more luminous than the Sun. It is about 5,100 light-years away and is expected to end its life as a supernova. The pulsation of the star has a period of about 740 days. It was first discovered and cataloged in 1860 by German astronomers at the Bonn Observatory. A hypothetical object traveling at the speed of light would take 14.5 seconds to orbit the Sun once. It would take that same object about 7 hours to orbit UY Scuti. [source: "UY Scuti Facts," *nineplanets.org*, Mar 6, 2020]

Venus. Venus spins on its axis clockwise, making it the only planet in the Solar System that spins in retrograde. There are more volcanoes on Venus than any other planet. Astronomers have discovered over 1,600 volcanoes on Venus so far. Scientists think that most of these volcanoes are dormant. A day on Venus lasts 343 Earth days, while a year on Venus is shorter, at just 225 days. The temperature on Venus can reach 870 degrees Fahrenheit due to an extreme greenhouse effect and the atmosphere being made up of mostly carbon dioxide. The air pressure on Venus is 90 times higher that the pressure at sea level on Earth, That's equivalent to about half a mile under the ocean. Winds can reach 450 miles an hour on Venus. The Venusian winds are faster than the strongest tornadoes on Earth.

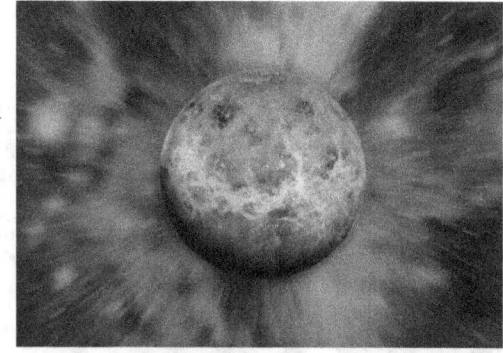

Oddities of Science

Oddities in Biology

Bacteria cells. Bacteria (singular bacterium) are a type of microscopic biological single-celled organism. The human body carries more bacteria cells than human cells. It used to be thought from a 1972 study that there were 10 times as many microbial bacteria cells in the human body as there are human cells. New studies now show that there are 30 trillion human cells (mostly red blood cells) and 39 trillion bacteria in the human. The highest numbers of bacteria are found in the gut. They aid in digestion, stave off colonization by harmful pathogens, and help to develop the immune system. At least 2 million people in the United States are infected with antibiotic-resistant bacteria every year, leading to the death of at least 23,000 people. [source: Abbott, "Scientist bust myth that our bodies have more bacteria than human cells," Nature, Jan 8, 2016]

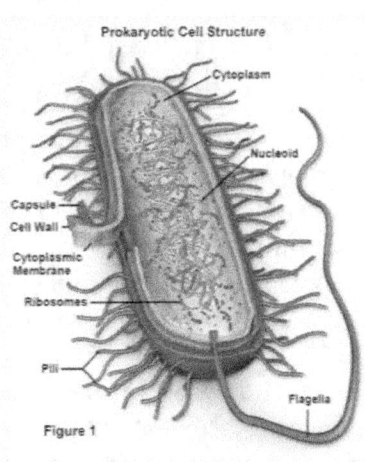

Figure 1

Brain. The human brain is the largest brain of all vertebrates relative to body size. The brain weighs about 3.3 pounds (a whale's brain weighs 20 pounds). The brain makes up about 2% of a human's body weight. The cerebrum makes up 85% of the brain's weight. The brain contains about 86 billion nerve cells (neurons). The fastest speed of information to pass between neurons is about 250 miles per hour. About 75% of the brain is made up of water. The human brain grows 3 times its size in the first year of life. [source: Lewis, "Human Brain: Facts, Functions & Anatomy," *Live Science*, Sep 28, 2018]

Butterfly. Many butterflies have circular eyespot patterns (called ocellus) on their wings that are used to deflect attacks from predators and attract mates. However, in dry seasons, butterflies try to avoid drawing attention to themselves by shrinking their eyespots to make them look like a dead leaf. In some species, the conspicuous eyespots are hidden at rest to decrease detectability, and only exposed when they believe potential predators are nearby. There are some 18,500 species of butterfly, found in every continent except Antarctica. The British painted lady butterfly migrates from tropical Africa to the Arctic Circle and back. That's a 9,000-mile round trip that takes up to 6 generations. Butterflies navigate using a time-compensated sun compass. They can see polarized light, which allows them to see and orient themselves even in cloudy conditions. [source: "How some butterflies developed the ability to change their eyespot size," *Science Daily*, Feb 11, 2020]

Chromosome. A chromosome is a thread-like DNA molecule with part or all of the genetic material of an organism. It carries the hereditary information for everything from height to eye color. In 1921, Theophilus Painter (1889-1969) gave the number 24 for the count of human chromosome pairs, for a total of 48 chromosomes. This number was undisputed for over 30 years. Then, in 1955, Joe Hin Tjio (1919-2001) was the first person to recognize the normal number of chromosomes as 46 rather than 48. Humans have 22 chromosome pairs and 2 sex chromosomes. Females have two X chromosomes and males have an X and a Y chromosome. [source: Szalay, "Chromosomes: Definition & Structure," *Live Science*, Dec 9, 2018]

Dolly. Dolly (1996-2003) was a female domestic sheep, and the first mammal to be artificially cloned from an adult somatic cell using the process of nuclear transfer. Somatic cell transfer means that the cloned nucleus comes from an adult and was placed in an egg cell from the same species. The single cell used as the donor for the cloning of Dolly was taken from a mammary gland of an adult sheep. One of the cloners stated that "Dolly is derived from a mammary gland cell and we couldn't think of a more impressive pair  of glands than Dolly Parton's." In 2013, Dolly (the sheep) was euthanized after developing a lung disease and severe arthritis. In 2000, the first pig was cloned. In 2001, the first cat and goat were cloned. In 2003, the first horse and rabbit were cloned. In 2005, the first dog was cloned. In 2009, the first camel was cloned. In 2010, the first buffalo was cloned. In 2011, the first coyote was cloned. [source: Gabbatiss, "Dolly the sheep: 15 years after her death, cloning still has the power to shock," *The Independent* (UK), Feb 14, 2018]

Genes. In biology, a gene is a small sequence of nucleotides in DNA or RNA that encodes the synthesis of a gene product such as RNA or protein. They contain all the instructions for our individual characteristic such as eye and hair color. Humans have about 25,000 genes. The word 'gene' was coined in 1909 by Danish botanist Wilhelm Johannsen (1857-1927). He described Mendel's individual units of heredity as genes, derived from pangenesis. Pangenesis was the word Charles Darwin used for the now-disproven theory of heredity. 99% of all genes are identical in every human. The most studied gene is called TP53, a gene that is a tumor suppressor, and known as the 'guardian of the genome.' It is mutated in half of all human cancers. [source: Dolgin, "The most popular genes in the human genome," *Nature*, Nov 22, 2017]

Great Barrier Reef. Australia's Great Barrier Reef is the largest living structure on Earth. It is the world's longest and largest coral reef system composed of about 3,000 individual reefs that stretches over 1,400 miles. The reef is about 6,000 to 8,000 years old. It is dying. The reef has lost more than half its coral since 1985 due to environmental pressures and polluting. In 2016, a 500-mile stretch of coral died due to high water temperature from the effects of global climate change. There have been over 1,600 known shipwrecks in the Great Barrier Reef region. In 2010, a coal carrier ship ran aground in the Reef region, spilling up to 4 tons of oil into the water and causing extensive damage to the reef. In 2015, more than 2 billion corals lived in the Great Barrier Reef. Now, half of them are dead. [source: Meyer, "Since 2016, Half of All Coral in the Great Barrier Reef Has Died," *The Atlantic*, Apr 18, 2018]

Hair. Hair is a tough protein filament that grows from follicles found in the dermis. Hair is one of the defining characteristics of mammals. Blondes have about 150,000 hairs on their scalp. Brunettes have about 100,000 hairs on their scalp. Each hair can support a weight of 2.5 ounces. This means that the combined strength of 150,000 hairs would be enough to support 12 tons. A rope made from 1,000 hairs could lift a full-grown

man. Your entire body, except for the palms of your hands and the soles of your feet, is covered with hair. A human being has about 5 million follicles, about the same as chimps and other primates. Hair grows faster than anything else in your body. Hair grows ½ inch a month. [sources: Hoffman, "Hair (Human Anatomy)," *WebMD*, May 18, 2019 and "How to be Human: The reason we are so scarily hairy," *New Scientist*, Oct 4, 2017]

Henneguya salmincola. The parasitic blob known as Henneguya salmincola is the only known animal on Earth that does not breathe. It spends its entire life infecting the muscle tissues of fish and underwater worms. There are no respiratory genes in this parasite. Scientists are not sure how the parasite acquires energy. [source: "A cnidarian parasite of salmon lacks a mitochondrial genome," *Proceedings of the National Academy of Sciences*, Jan 10, 2020]

Jellyfish. The jellyfish known as Turritopsisdohrnii is an odd species. It is a biologically immortal species. It starts life as a tiny free-swimming larva known as planulae, and then it lives for a while as a polyp on a hard surface that eventually gives rise to an adult jellyfish. The polyps form into an extensively branched form, which is not commonly seen in most jellyfish. These adult jellyfish can revert back from being a sexually mature adult to being an immature polyp and begin the process again. Most other jellyfish species have a relatively fixed life-span, which varies by species from hours to

many months. The Turritopsisdohrnii is the only form known to have developed the ability to return to a polyp state and live on forever if not eaten or become diseased. The numbers of jellyfish are growing. By clogging cooling equipment, jellyfish have shut down nuclear power plants. In 2006, jellyfish partially disabled the aircraft carrier *USS Ronald Reagan*. Jellyfish kill about 100 people a year with its

Oddities of Science

sting. [source: Tucker, "Jellyfish: The Next King of the Sea," *Smithsonian Magazine*, August 2010]

Junk DNA. In genetics, the term junk DNA refers to regions of DNA that are conceding. It turns out that 98% of the human DNA is junk DNA, while in bacteria, only 2% of the genetic material is junk DNA that does not code for anything. The term "junk DNA" was first used in the 1960s and was formalized by Susumu Ohno (1928-2000). It was he who popularize the term junk DNA for segments of the DNA that have no functions. He had noticed that the amount of mutation occurring as a result of deleterious mutations set a limit for the amount of function segments that could be expected when a normal mutation rate was considered. It turns out that some of the junk DNA act like a conductor influencing the pace and repetitions of the coding sequence. Scientists have now linked various non-coding sequences to various biological processes and even human diseases. Junk DNA is also the portion employed in DNA fingerprinting. This was first used in 1986 when British geneticist Alec Jeffreys (1950-) was asked to help in a disputed immigration case to confirm the identity of a British boy whose family was originally from Ghana. The DNA results proved that the boy was closely related to the other members of the family. The police then asked if he could help identify the killer of two female teenagers. The murderer was identified and convicted of their murders after samples taken from him matched semen samples from the dead girls. Despite the number of functions now described to junk DNA, many geneticists think that at least 75% of it has no function. [source: Blanco, "Our Cells Are Filled With 'Junk DNA' – Here's Why We Need It," *Discover Magazine*, Aug 13, 2019]

Mesentery. In 2017, biologists officially added a new organ to Gray's Anatomy. The new organ, called the Mesentery, is now officially the human body's 79th organ. It is a double-folded organ that attaches the intestines to the abdominal wall in humans. The Mesentery was originally thought to be a fragmented structure that was part of the digestive system. However, scientists have now discovered that it is one continuous organ. It was first identified by Dr. Calvin Coffey in Limerick, Ireland in 2016. The new organ's function is still a mystery. [sources: MacDonald, "It's Official: A Brand-New Human Organ Has Been Classified, *Science Alert*, Jan 3, 2017 and Coffey, "The mesentery: structure, function, and role in disease,: *The Lancet*, Vol 1, Issue 3, Nov 1, 2016]

Nose. The human nose is the first organ of the respiratory system and is the principal organ of the olfactory system (sense of smell). The main function of the nose is respiration. Another main function is the sense of smell. The upper nasal cavity, which contains millions of specialized olfactory cells, is responsible for this function. Humans have about 450 different types of olfactory receptors (dogs have about 900 receptors). The nose can outperform the eyes and ears. The eyes can discriminate about 2 million colors. The ears can discriminate about half a million tones. But the nose can discriminate about a trillion smells. One odd fact  about the nose is that you cannot smell while you are asleep. The olfactory sensory organs shut down during **REM** sleep. That's why smoke and fire alarms are so important in the home. You cannot smell the smoke and wake up. Some people do dream that they smell something, but that isn't due to the nose. That has to do with your subconscious. The largest nose on a living person belongs to Mehmet Ozyurek of Turkey. His nose is 3.46 inches long from bridge to the tip. Every day your nose cleans, moistens, and warms more than 18 cubic yards of air. When you breathe gently, air passes through the nostrils at the rate of 2 yards a second. [source: Bradford, "Nose: Facts, Function & Diseases," *Live Science*, Sep 30, 2015]

Redwood trees. Redwood trees (sequiodeae) are the largest and tallest trees in the world. Oddly, they grow only one place on Earth, in an area from Big Sur, northern California to southern Oregon, a distance of 450 miles along the western slopes of the Sierra Nevada range. They outgrow all other trees and can live to be over 2,200 years old. The tallest tree in the world is a Redwood tree 380.1 feet tall with a width of over 22 feet. The Redwood tree grows from a cone only one inch long and contains about 50 to 60 seeds. Redwood trees have been around for about 240 million years. Redwood trees capture more carbon dioxide (CO_2) than any tree on Earth. A redwood's bark can be one foot thick. It contains tannin which protects the tree from fire, insects, fungus, and diseases. [source: "About the Trees: The Redwood Tree," *nps.gov*]

Oddities of Science

Tardigrades. Tardigrades (meaning "slow steppers"), also known as water bears or moss piglets, are odd creatures that can survive extreme conditions. They have been found everywhere, from the highest mountaintops to the deep sea and mud volcanoes, from tropical rainforests to the Antarctic. They look like microscopic, 8-legged bears. They are among the most resilient animals known. On Earth, they have been exposed to extreme pressures, large radiation doses, frozen for decades without food or water, then springing back to life when unfrozen. In 2007, a number of dried up tardigrades were sent into space. They were exposed to the cold vacuum of space, solar winds and the sun's radiation for 10 days, and then brought back to earth. The tardigrades exposed to more radiation were damaged (ultraviolet radiation damaged the cellular material and the DNA), but all of them survived, and many went on to lay eggs that successfully hatched back on Earth. Before this experiment, only lichen and bacteria were known to be able to survive exposure to the combination of vacuum and space radiation. [source: Courtland, "Water bears are first animal to survive space vacuum," *New Scientist*, Sep 8, 2008]

Telomere. A telomere is a region of repetitive nucleotide sequences at each end of a chromosome. It protects the end of a chromosome from deterioration or from fusion with neighboring chromosomes. Telomeres help to organize each of the human's 46 chromosomes in the nucleus. Without telomeres, important DNA would be lost every time a cell divides (about 70 times in a lifetime). The length of telomeres decreases with age. In 1975, Elizabeth Blackburn (1948-) discovered the odd nature of telomeres. She and Carol Greider and Jack Szostak were awarded the 2009 Nobel Prize in Physiology or Medicine for the discovery of how chromosomes are protected by telomeres. [source: "What is a telomere?" *yourgenome.org*, Jan 25, 2016]

Treehoppers. Treehoppers are a group of insects that have bizarre protuberances that look like horns, gnarled branches, antlers, brightly colored flags, or dead plant leaves. They can't help it. They were born this way. They suck on plant juices. There are about 3,200 species of treehoppers. They are one of the most diverse bug groups on the planet. Individual treehoppers live for only a few months. They sing to each other by vibrating plant stems. In 2020, Brendan Morris, a graduate student at the University of Illinois at Urbana-Champaign, discovered a new species and named it after Lady Gaga, calling it Kaikaia gaga. [source: Hegedus, "Insect named after Lady Gaga," *New York Post*, March 11, 2020]

Venus Flytrap. The Venus Flytrap (Dionaea muscipula) is an odd carnivorous plant that grows in the Carolinas. The trap is triggered by trigger hairs, called trichomes. When two hairs are touched within 20 seconds of each other by an insect or spider or, if a single hair is touched twice, the trap snaps shut. If the prey is too small and escapes, the trap will usually reopen within 12 hours. If the prey moves around in the trap, it tightens and digestion begins more quickly. Flytraps lure insects by the reddish lining in the leaves and by secreting fragrant nectar. The species is currently under Endangered Species Act review.  [source: Bradford, "Facts About Venus Flytraps," *Live Science*, Feb 25, 2017]

Yawn. A yawn is a reflex consisting of the simultaneous inhalation of air and the stretching of the eardrums, followed by an exhalation of breath. Scientists still aren't sure why we yawn. In humans, yawning is often triggered by the perception that others are yawning. This "contagious" yawning has also been observed in chimps, dogs, cats, birds, and reptiles. Some scientists think that the idea of yawning is a thermoregulatory behavior that cools down the brain, but its true biological function is still unclear. Yawning has been observed in fetuses at 30 weeks of pregnancy. [source: "Why do we yawn when we are tired? And why does it seem to be contagious?" *Scientific American*, March 20, 2002]

Oddities of Science
Oddities in Chemistry

Aerogel. Aerogel is a synthetic porous ultra light material derived from a gel, in which the liquid component for the gel has been replaced with a gas. It Is the lightest solid known to man. It was invented in 1931 by American scientist Samuel Kistler (1900-1975) as a result of a bet on who could replace the liquid in "jellies" with gas without causing any shrinkage. It is 99.8% air and one of the best thermal insulators. 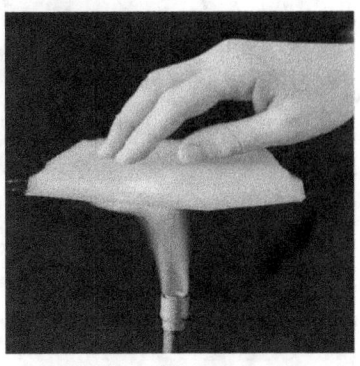 It has been used in NASA's *Stardust* space mission to collect dust from a comet's tail. [source: Calderone, "4 unbelievable chemical substances humans have discovered," *Business Insider*, July 9, 2015]

Aspirin. Aspirin is a medication used to reduce pain, fever, or inflammation. It can also be used as a blood thinner. It is based on a natural chemical found in willow bark. In 1853, chemist Charles Gerhardt (1816-1856) treated the medicine sodium salicylate with acetyl chloride to produced acetylsalicylic acid (ASA) for the first time. That is the main ingredient of aspirin. In 1899, scientists at the German pharmaceutical company Bayer company began studying ASA as a less-irritating medication, which they called "Aspirin." It is estimated that 44,000 tons, or 120 billion pills, of Aspirin is consumed every year. [source: Brazier, "Uses, benefits, and risks of aspirin," *Medical News Today*, Dec 18, 2017]

Astatine. The rarest naturally occurring element is astatine (element 85, abbreviated At). It is named after the Greek word for unstable (astatos). It is a radioactive product produced from the decay of uranium and thorium. It has a half-life of only 8.1 hours. A sample of the pure element has never been assembled since any specimen would be immediately vaporized by the heat of its own radioactivity. It was first synthesized in 1940. Less than 1 gram of Astatine is present at any given time in the Earth's crust. [source: Ross, "Facts About Astatine," *Live Science*, May 24, 2017]

Bakelite. Bakelite was the first plastic made from synthetic components. It was first developed by Belgian-American chemist Leo Baekeland (1863-1944) in 1907. It was patented in 1909. In all, Baekeland took out more than 400 patents related to the manufacture and applications of Bakelite. Bakelite was an inexpensive, non-flammable and versatile plastic, which marked the beginning of the modern plastics industry. It was soon used to make electrical insulators, radio and telephone casings, kitchenware, jewelry, pipe stems, children's toys, and firearms. [source: "Leo Hendrick Baekeland and the Invention of Bakelite," *American Chemical Society*]

Francium. Francium (element 87) is an extremely radioactive element, with a half-life of only 22 minutes. It is the second rarest naturally occurring element (after astatine). Bulk francium has never been viewed. The extreme heat of decay caused by its short half-life would immediately vaporize any viewable quantity of the element. It was discovered in 1939 by Marguerite Perey (1909-1975) of France, hence its name. She was a student of Marie Curie and later died of radiation poisoning. Francium was the last element first discovered in nature, rather than by synthesis. As little as 30 grams (1 ounce) exists at any given time throughout the Earth's crust. The largest amount produced in the laboratory was a cluster of about 300,000 atoms. [source: Greenwood, "My Great-Great-Aunt Discovered Francium, And it Killed Her," *New York Times*, Dec 7, 2014]

Glucose. Glucose (blood sugar) is a simple sugar and the most abundant monosaccharide, a subcategory of carbohydrates. Glucose is one of the most well-known molecules. Glucose is the most important source of energy in all organisms. The name glucose derives from the Greek "glukos," meaning "sweet." It was first isolated from raisins in 1747. Glucose was discovered in grapes in 1792. Glucose is made by plants using photosynthesis. Emil Fischer (1852-1919) studied how sugars worked and how glucose was formed. It earned him a Nobel Prize in Chemistry in 1902. For the discovery of the metabolism of glucose, Otto Meyerhof (1884-1951) received the Nobel Prize in Physiology or Medicine in 1922. [source: "Everything You Need to Know About Glucose, *Healthline*, March 24, 2017]

Graphite. Graphite is a crystalline form of the element carbon. Under high pressures and temperatures, it converts to diamond. Graphite is used in pencils and lubricants. There is enough graphite in the human body to make 9,000 pencils. Graphite is a good conductor of heat and electricity.

Oddities of Science

Its high conductivity is useful in making electrodes, batteries, and solar panels. It is the only non-metal that can conduct electricity, but acts as a nonmetal that resists high temperature. In 1789, Abraham Werner named graphite for its ability to leave marks on paper and other objects. It is derived from the Greek word "graphein" meaning "to write or draw." In 1795, the modern pencil that uses graphite was invented by **Nicholas Conte**. [source: Kielmas, "What Are the Uses of Graphite," *Sciencing*, April 5, 2018]

Helium. Helium was first detected in the Sun before it was detected on Earth. It was detected in 1868 during a solar eclipse using spectral lines. Helium wasn't discovered on Earth until 1895 when it was

found that helium was emanating from uranium ore. In 1903, residents of Dexter, Kansas stumbled across a geyser of gas that didn't burn. The gas was later found to contain a small percentage of helium. Unlike any other element, helium will remain a liquid all the way down to absolute zero (-273 degrees Fahrenheit). Superfluid helium has the odd property of defying gravity and can climb on walls. It is known as the creeping effect. When liquid helium approaches absolute zero, part of the liquid becomes a "superfluid." It has zero viscosity which will move rapidly through any pore in an apparatus. Superfluid helium will creep along the sides of a container until it reaches a warmer region where it evaporates. Helium's boiling point is the lowest among all the elements. [source: Greshko, "We Discovered Helium 150 Years Ago. Are We Running Out?" *National Geographic*, Aug 17, 2018]

Keratin. Keratin is a protective protein that is the key structural making of hair, nails, feathers, horns, claws, hooves, and calluses. Keratin can also be found in your internal organs and glands. Certain fungi eat keratin, which is why it can be hard to get rid of athlete's foot. [source: "Keratin – an overview," *Science Direct*]

LSD. Lysergic acid diethylamide (LSD) is an illegal hallucinogenic drug that alters the senses and cause hallucinations. LSD was not invented as a way for people to get high. In fact, its psychedelic properties were not even discovered until 5 years later. In 1938, Albert Hoffman (1906-2008) was looking for alkaloid derivatives. He had been researching lysergic acid, a chemical from ergot, a fungus found in tainted breads. He synthesized it to obtain a respiratory and circulatory stimulant. Five years later, Hoffman decided to work with lysergic acid some more. He accidently absorbed a small amount of the drug through his fingertips. He soon grew dizzy and had to take a break from working on it. In his journal, he described a feeling as though he was drunk, but with a highly stimulated imagination. He described an uninterrupted stream of weird pictures, extraordinary shapes, and intense, kaleidoscopic play of colors. A few days later, Hoffman ingested some LSD and started to feel the psychedelic effects of the drug as he rode home on a bike. This was the first intentional LSD trip. In the 1950s, the CIA believed that the drug might be useful for mind control. It currently has no approved medical use. [source: Davis, "The effects and hazards of LSD," *Medical News Today*, June 22, 2017]

Osmium. Osmium (from the Greek word for smell) is a chemical element with the symbol Os and an atomic weight of 76. It was discovered, along with iridium, by Dr. Smithson Tennant (1761-1815) when extracted large amounts of platinum. It is the densest naturally occurring element in the world. It is the rarest of the stable elements. There is 1 gram of osmium per 200 tons of the Earth's crust. The worldwide production of osmium is 500 kilograms, which is 5,000 times less than its gold counterpart. It is used to make pen nib tipping, electrical contacts, and other applications that require extreme durability and hardness. [source: Girolami, "Osmium weighs in," *Nature*, Oct 23, 2012]

Ozone. Ozone is an inorganic molecule with the chemical formula O_3. It is a pale blue gas that smells like chlorine. It can be formed by the action of ultraviolet light (UV) and electrical charges, such as lightning. The ozone layer (about 6 to 31 miles above the surface) in the atmosphere absorbs most of the harmful UV rays from the Sun. In 1839, Christian Schoenbein succeeded in isolating the gas and named it ozone, from the Greek word ozein, meaning "to smell." The largest use of ozone is in the preparation of pharmaceuticals and disinfectant. Many municipal drinking water systems kill bacteria

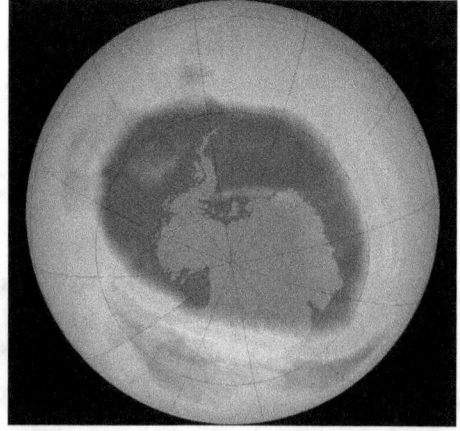

Oddities of Science

with ozone instead of chlorine. Ozone does not leave any taste or odor in drinking water. In 1985, scientists documented the thinning of the ozone layer above Antarctica. Now, the "hole" in the stratosphere above Antarctica is the smallest in three decades. The better-than-expected levels of ozone have been attributed to a sudden warming at high altitudes. It was also helped by the dramatic reduction in chlorofluorocarbons (CFCs), the chemicals that caused the problem in the first place. An international treaty helped phase out CFCs and most of the chlorine- and bromine-containing chemicals involved in ozone depletion. [source: Amos, "Ozone hole vigilance still required," *BBC News*, Oct 8, 2019]

Penicillin. In 1928, Dr. Alexander Fleming (1881-1955) of Scotland accidently discovered the world's first antibiotic substance, penicillin, from mold. In his London lab, after returning from a vacation, he observed how a mold growing on one of his Petri dishes he accidently left open suppressed the growth of nearby bacteria. But, despite his best efforts, he failed to extract any usable penicillin (which he called mold juice), and he gave up. His work met with little interest and even less enthusiasm by his peers. In 1939, Australian pharmacologist Howard Florey (1898-1968) figured out a way of purifying penicillin in useable quantities. He carried out the first ever clinical trial of penicillin in

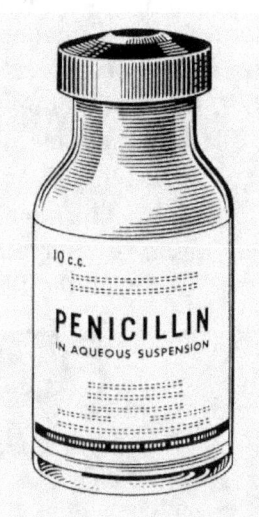

1941 with the help of Dr. Joseph Chain. Florey's first patient, a policeman named Albert Alexander, started to recover from a face infection, but he died because Florey could not make enough penicillin. In December 1942, survivors of the Cocoanut Grove fire in Boston were the first burn patients to be successfully treated with penicillin. In 1944, chemical engineer Margaret Hutchinson Rousseau (1910-2000) designed the first commercial penicillin plant. She developed a deep-tank fermentation of penicillium mold that enable large-scale production of penicillin. She was the first female member of the American Institute of Chemical Engineers. Less than 1% of people are dangerously allergic to penicillin. [source: Newman, "How do penicillins work?" *Medical News Today*, July 30, 2018]

Periodic Table. The periodic table is a tabular display of the chemical elements. It is arranged by atomic number, electron configuration, and recurring chemical properties. The first element, with the atomic number of 1, is hydrogen (H). The element with the largest atomic element, 118, is oganesson (Og). The first 94 elements all occur naturally. Elements 95 to 118 have only been synthesized in laboratories or nuclear reactors. In 1939, the French chemist Marguerite Perey discovered an element of her own. She wanted to call it "catium" to honor the particle's strong attraction to cathodes. Marie Curie's daughter, Irene, worried that English speakers would associate the element with house cats. Perey, being French, decided to call it francium instead. In 1962, Alexander de Chancourtois published an early form of the periodic table. He was the first person to notice the periodicity of the elements. In 1869, Russian chemist Dmitri Mendeleev published the first recognizable periodic table. When Mendeleev produced his first table, only 65 elements were known. In 1939, the first element synthesized was Neptunium (Np), element 93. In 2010, the last element to be synthesized was Tennessine (Ts), element 117. The joint Russia-US collaboration was able to synthesize 6 atoms of it. The race is on to find element 119. The only letter not appearing on the periodic table is the letter J. [source: Jahromi, "The Histories Hidden in the Periodic Table," *The New Yorker*, Dec 27, 2019]

Phosphorus. Phosphorus, the 15th element on the periodic table, is a chemical element with the symbol P that is found in pee. It was the first element to be discovered that was not known since ancient times. It was discovered by accident. In 1669, the German alchemist Hennig Brand (1630-1710) became convinced that gold could be distilled from urine as he hunted for the "philosopher's stone." He assembled 60 buckets of human urine, which he kept for months in his cellar, which gave off a terrible smell. He converted it to a paste, and then to a translucent waxy white substance. No gold was found, but he found the substance glowing, and when it was exposed to air, it often spontaneously burst into flame. He put some of the liquid in a jar and covered it, where it solidified, and continued to give off a pale-green glow. Brand kept his discovery secret for a while, working with

Oddities of Science

the phosphorus trying unsuccessfully to use it to produce gold. He then sold his recipe to make phosphorus to D. Kraft from Dresden. In 1680, Robert Boyle (1627-1691) of England, the first modern chemist, found out and used phosphorus to ignite sulfur-tipped wooden splints, forerunners of our modern matches. In the 1750s, Carl Scheele, a Swedish chemist who discovered 8 elements (and got credit for none of them), devised a way to manufacture phosphorus without a lot of urine. In 1830, Charles Sauria invented the first striking match with a phosphorus head. [source: "Facts About Phosphorus," *Live Science*, Jan 6, 2015]

Quasicrystal. A quasiperiodic crystal, or quasicrystal, is an odd structure that is ordered, but never repeats themselves. This means that a shifted copy will never match exactly with its original. It has been called nature's impossible matter with an "impossible" atomic arrangement. It was discovered in 1982 by Dan Schechtman, an Iowa State professor of materials science and engineering. It took him two years to publish his results. In 1992, the International Union of Crystallography had to redefine just what was meant by a crystal. In 2011, Schechtman was awarded the Nobel Prize in Chemistry for his discovery. [source: Oberhaus, "Quasicrystals Are Nature's Impossible Matter," *vice.com*, May 3, 2015]

Ricin. Ricin is a carbohydrate-binding protein produced in the seeds of the castor oil plant, *Ricinus communis*. It is a highly potent toxin that inhibits protein synthesis. A dose of ricin the size of a few grains of table salt can kill a man. The seeds of *Ricinus communis* are commonly crushed to extract castor oil. As ricin is not oil-soluble, little is found in the extracted castor oil. 5 to 20 castor beans ingested can be fatal. In 2013, a 37-year-old woman in the USA survived after eating 30 castor beans. During World War I, ricin was investigated as a toxic dust or as a coating for bullets and shrapnel. In 1978, Bulgarian dissident Georgi Markov (1929-1978) died after a ricin pellet was fired from an umbrella into his right thigh in London. He metal pellet was the size of the head of a pin. The pellet was actually a jeweler's watch bearing, used in precision watch making. It was the first assassination by ricin poisoning. [source: "Umbrella Assassin: Clues and Evidence, *PBS*, June 3, 2014]

Urea. Urea, also known as carbamide, is an organic compound. It is the main nitrogen-containing substance in the urine of mammals. It is a colorless, odorless solid that is highly soluble in water. Formed in the liver, the body uses it for nitrogen excretion. Urea is widely used in fertilizers as a source of nitrogen. Its preparation by German chemist Friedrich Woehler (1800-1882) from ammonium cyanate in 1828 was the first lab synthesis of a naturally occurring organic compound from inorganic materials.

Water. Water (H_2O), also known as dihydrogen monoxide, is an odd inorganic molecule. Typically, when something is cold, it shrinks. That's because temperature is a measure of atomic vibration. The more vibration in a substance, the more space it takes, the more it expands. But water is an exception. Even though water vibrates less when it is frozen and becomes ice, the ice occupies more volume due to the strange shape of the water molecule. When water freezes, it releases energy because a lot of extra strong bonds can be made. This takes up more space. And so, ice expands when it freezes. An ice cube takes up 9% more volume than the water used to make it. Also odd is that hot water freezes faster than cold water. Water covers 71% of the Earth's surface. Water is the only known substance to exist as a solid, liquid, and gas in normal terrestrial conditions. One bucket of water contains more atoms than there are bucketfuls of water in the Atlantic Ocean. When light passes through water, its protons are only traveling at ¾ speed of light in a vacuum. There are about 332 million cubic miles of water in the world. 93% of this water is seawater.

Oddities of Science

Oddities in Computer Science

Computer Mouse. In 1963, the first computer mouse was invented by Douglas Engelbart (1925-2013) at Stanford Research Institute (now SRI International) in Menlo Park, with the help of Bill English (1938-). English built the first mouse, made of wood with 2 metal wheels, based on Engelbart's notes. Englebart (SRI) applied for a patent of the computer mouse in 1967 and received it in 1970. He first called it the "X-Y position indicator for a display system," or the "bug." He later nicknamed his device a "mouse" because the early models had a cord attached to the rear part of the device which looked like a tail and in turn resembled the common mouse. In December 1968, the first public demonstration of a mouse controlling a computer system was demonstrated at The Mother of All Demos computer conference in San Francisco. He never received any royalties for the invention of the mouse. SRI patented the mouse, but had no idea of its use or value. SRI later licensed it to Apple Computer for $40,000. The patent expired before the mouse became widely used in personal computers. English left SRI and went to Xerox PARC, where he developed the ball mouse, in which a ball replaced the original set of wheels. [source: "The History of the Computer Mouse," *computinghistory.org*]

Cray-1. The Cray-1 was a supercomputer designed, manufactured and marketed by Cray Research under the leadership of Seymour Cray (1925-1996). The first Cray-1 was installed at Los Alamos National Laboratory in 1976. Eventually, over 100 Cray-1 supercomputers were sold, making it one of the most successful supercomputers in history. The Cray-1 was the first computer to implement vector processor design. This implemented an instruction set containing instruction that operate on arrays of data called vectors, compared to scalar processors, whose instructions operate on single data items. The cost of a Cray-1 was $9 million (equivalent to $33 million today). Its speed was 160 million floating operations per second (MFLOPS). It weighed over 5 tons and needed 250 kilowatts of power for calculations, storage, and coolant. [source: Modine, "Remembering the Cray-1," *The Register*, Jan 5, 2008]

CTRL ALT DEL. The CTRL-ALT-DEL (Control-Alt-Delete) combination is a famous keyboard shortcut to soft reboot or bring a Close Program dialog on Microsoft Windows. The idea for soft rebooting via a keyboard came from Dr. David Bradley (1949-), a computer scientist and electrical engineer at IBM. He chose this shortcut for safety reasons to avoid accidental trigger. It was chosen because it was impossible to hold this combination with one hand on an IBM keyboard. The Ctrl-Alt-Del was originally not designed to be used by end users. It was meant to be used by people writing computer programs or documentation, so that they could reboot their computers without powering them down. Since software developers and technical writers would need to restart a computer many times, this key combination was a big time-saver. The key combination was supposed to be a secret, but the combination was described in IBM's technical reference documentation and thereby revealed it to the general public.

ENIAC. ENIAC (Electronic Numerical Integrator and Computer) was the first electronic general-purpose digital computer. What distinguished ENIAC from other computing machines was that it could be reprogrammed for different tasks. It was originally designed in secret (code name "Project PX") to calculate artillery firing tables for the Army, but its first program was a study of the feasibility of a thermonuclear weapon (hydrogen bomb) in November 1945. The input/output for this test used 1 million punch cards. The machine was located at the University of Pennsylvania. ENIAC could calculate a trajectory in 30 seconds that took a human 20 hours. The cost of the machine was about $500,000 (about $7 million today). It had 20,000 vacuum tubes, 70,000 resisters, 10,000 capacitors, 6,000 switches, and 5 million hand-soldered joints. It weighed 30 tons, took up 1,800 square feet, and consumed 150 kilowatts of electricity. It could do 5,000 operations a second. Vacuum tubes burned out almost every day, leaving ENIAC nonfunctional about half the time. In 1947, the machine was transferred to the Aberdeen Proving Ground, Maryland. It was in operation until October 1955. ENIAC was a one-of-a-kind design and was never repeated. The original designers, John Mauchly (1907-1980) and Presper Eckert (1919-1995), started a new computer firm (EMCC), but the firm struggled and the pair sold the company to Sperry Rand. Mauchly and Eckert patented their ENIAC design, but it was invalidated by the US Patent Office in 1973. Eckert went on to design the first commercial computer, the UNIVAC I. [source: Levy, "The Brief History of the ENIAC Computer," *Smithsonian Magazine*, November 2013]

Ferranti Mark 1. In February 1951, Ferranti delivered their first Mark 1 computer to the University of Manchester. It was the world's first commercially available general-purpose electronic digital computer. The computer was a few months ahead of the UNIVAC I,

which was sold to the United States Census Bureau in March 1951, but not delivered until December 1952. It stored data in cathode-ray tubes and a rotating magnetic drum. The Mark I computer had 4,050 vacuum tubes. The second Mark I was supposed to go to the British Atomic Energy Research Establishment in the autumn of 1952, but a change in government canceled the sale. Ferranti ultimately sold it to the University of Toronto. The Mark 1 was used to help forecast election results, calculate wages, and produce actuarial tables, among other things. In 1954, Alan Turing (1912-1954) devised a set of rules for a chess program, which Turing started to code for the Ferranti Mark 1, before his untimely death. [source: Ackerman, "The Ferranti Mark 1: World's First Commercially Available General-Purpose Computer," *IEEE Spectrum*, April 29, 2016]

K computer. The K computer, named for the Japanese word/numeral "kei," was a supercomputer manufactured by Fujitsu. It had over 80,000 CPUs or computer nodes and over 200,000 cables. The development costs after 6 years of development was $1 billion. In June 2011, it was ranked as the world's faster computer, with a computational speed of over 8 petaflops (8 quadrillion operations per second). In November 2011, it became the first computer to top 10 petaflops. In June 2012, K was superseded as the world's fastest computer by the American IBM Sequoia at Lawrence Livermore Laboratory. Its speed was 20.13 petaflops. In August 2019, the K computer was decommissioned. It had been used in weather forecasting, semiconductor development, and medical research. It will be succeeded by the Fugaku, expected to be 100 times faster. It is scheduled to be operation in 2022. [source: "Japan's L supercomputer, once the world's fasters, to retire in August, *Japan Times*, Feb 6, 2019]

Lisp. Lisp (LISt Processor) is a family of computer programming languages. It was originally specified by John McCarthy (1927-2011) in 1958. It is the second-oldest high-level programming language in widespread use today. Only Fortran is older, created in 1957. Lisp was originally created as a practical mathematical notation for computer programs. It quickly became the favored programming language for artificial intelligence (AI) research (McCarthy coined "Artificial Intelligence") after its publication in 1960. It was called a LISt processor because linked lists are one of Lisp's major data structures, and Lisp source code is made of lists. [source: "How Lisp Became God's Own Programming Language," *twobithistory.org*, Oct 14, 2018]

Nanoblob. Computer scientists have taught a thin film of nanomaterials (called a nanoblob) to solve classification problems, such as spotting a cancerous lesion in a mammogram. The scientists used evolutionary algorithms and a custom circuit board to send voltage pulses through an array of electrodes into a dilute mix of carbon nanotubes dispersed in a liquid crystal. Over time, the carbon nanotubes arranged themselves into a complex network that spanned the electrodes. The network was then capable of carrying out the key part of an optimization problem. The blob could then learn to solve a second problem. [source: Moore, "4 Strange New Ways to Compute," *IEEE Spectrum*, Nov 29, 2017]

RAMAC. The IBM 305 RAMAC (Random Access Method for Accounting and Control) was the first commercial computer that used a moving-head hard disk drive (HDD) for secondary storage. It was launched in September 1956, was 16 square feet in size, weighed over 2,000 pounds, and could store 5 megabytes (MB) of data (equivalent to 64,000 punched cards). It could be leased from IBM for $3,200 a month (equivalent to $29,000 in to-

Oddities of Science

day's currency). It was discontinued in 1961. Over 1,000 RAMACs were built. Its first use was in the U.S. auto industry for inventory control. During the 1960 Olympic Winter Games in Squaw Valley, California, it was used to provide the first electronic data processing systems for the Games. The RAMAC was one of the last vacuum tube computers. The storage capacity could have been greatly increased beyond 5 MB, but IBM did not know how to sell a product with more storage. [source: VanHemert, "IBM 305 RAMAC: The Grandaddy of Modern Hard Drives," gizmodo.com, Mar 16, 2010]

Robot. In January 1979, Robert Williams (1953-1979) was the first person to die by a robot. A computer-programmed parts retrieval robotic arm collided with him while it was trying to fetch a part. Williams was an American factory worker at the Ford Motor Company in Michigan. Williams had climbed into a storage rack when he was struck from behind by a robotic arm and killed instantly. His body remained in the shelf for 30 minutes until he was discovered by workers who were concerned about his disappearance. Williams' family won a $10 million lawsuit for his wrongful death. In 1981, Jenji Urada, age 37, died when he failed to power down a robot he was working on. The robot's hydraulic arm shoved him into a grinding machine. [source: Young, "The First 'Killer Robot' Was Around Back in 1979, " *How Stuff Works*, April 9, 2018]

RSA Cryptosystem. RSA (Rivest-Shamir-Adleman) is one of the first public-key cryptosystems. It is used to secure data transmission. The encryption key is public and is different from the decryption key, which is kept secret (the private key – a large primary number). The RSA algorithm is based on the difficulty of the factorization of the product of two large prime numbers. The acronym is made for the MIT inventors of the algorithm, Ron Rivest (1947-), Adi

Shamir (1952-), and Leonard Adleman (1945-). They first published a paper describing the algorithm in 1977. In 1973, Clifford Cocks (1950-), a British mathematician and cryptographer, while working for the British intelligence agency Government Communications Headquarters (GCHQ), invented a similar algorithm, but his work

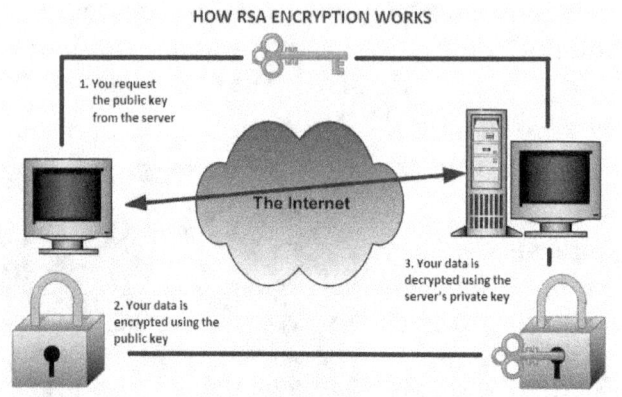

was considered top secret and was not declassified until 1997. Given the relatively expensive computers at the time needed to implement the algorithm, the UK government considered it mostly a curiosity and never deployed it. There are no currently published methods to crack the RSA system if a large enough key is used. RSA is a relatively slow algorithm, and as a result, it is less commonly used to directly encrypt user data. In 1983, MIT patented the cryptographic communications system and method that used the algorithm. In 2000, the detailed description of the algorithm was published two weeks before the end of the patent. Had Cock's work been publicly known, that patent on the RSA algorithm and cryptographic communications systems would have been illegal. In 2019, French scientists set a world record by extending the longest cracked encryption from 232 digits to 240. Fast computers work simultaneously around the world used 35 million hours of computing time to do the job. The cryptographic key was 795 bits long, but most computers use the 2048-bit encryption. [source: Delbert, "Scientists Crack Longest, Most Complex Encryption Key Ever," *Popular Mechanics*, Dec 6, 2019]

Software bug. A software bug (or computer bug) is an error or flaw in a computer program or system that causes an incorrect or unexpected result. The process of finding and fixing bugs is termed "debugging." In September 1947, a dead moth was removed from the Harvard Mark II

computer (Aiken Relay Calculator). It was not found by Grace Hopper as some sources say. The moth was trapped in the number 70 relay, Panel F, coining the term "bug." The moth was later taped to the computer's logbook, with the notation "First actual case of bug being found." In 1996, the European Space Agency (ESA) launched an Arianne 5 rocket that self-destructed less that 1 minute after launch due to software bugs in its guidance systems. [source: Van Zanten, "The very first recorded computer bug," *The Next Web*, Sep 18, 2013]

Therac-25. The Therac-25 was a computer-controlled radiation therapy machine ("cancer zapper") produced by the Atomic Energy of Canada Limited (AECL) in 1982. From 1985 through 1987, the Therac-25 was involved in at least 6 medical accidents, in which patients were given massive overdoses of radiation. Several days later, radiation burns appeared. In four cases, the injured patients died as a result of the overdose. The other two were left with lifelong injuries. Programming errors (software bugs) gave the patients radiation doses that were hundreds of times greater than normal. There were no hardware interlocks or supervisory circuits to ensure that software bugs couldn't result in catastrophic failures. The accidents highlighted the dangers of software control of safety-critical systems. [source: Fabio, "Killed by a Machine: The Therac-25," *Hackaday,* Oct 26, 2015]

Turing Award. The Association for Computing Machinery (ACM) Turing Award is an annual prized given to an individual selected for contributions "of lasting and major technical importance to the computer field." It is recognized as the highest distinction in computer science and is called "the Nobel Prize of computing." The award is named after Alan Turing (1912-1954), considered the key founder of theoretical computer science. In 1966, the first recipient was Alan Perlis (1922-1990) for his influence in the area of advanced computer programming techniques and compiler construction. Starting in 2007, the award was accompanied by an additional prize of $25,000. Since, 2004, the award has been accompanied by a prize of $1 million, with financial support provided by Google. In 2019, the award was won by three pioneers in Artificial Intelligence – Yann Lecun, Geoffrey Hinton, and Yoshua Bengin. All three worked on the development of neural networks. [source: Metz, "Turing Award

Won by 3 Pioneers in Artificial Intelligence," *New York Times*, March 27, 2019]

UNIVAC. The UNIVAC 1103, also known as Atlas II or ERA 1103, was the first commercially successful scientific computer. The computer was designed by Seymour Cray (1925-1996) at Engineering Research Associates (ERA) and built by Remington Rand in October 1953. The computer weighed 38,543 pounds using vacuum tubes. The first UNIVAC 1103 was sold to aircraft manufacturer Convair.

Y2K Fiasco. The Y2K fiasco, also known as the Millennium Bug, was the result of 20th century programmers allotting just two digits to register years, failing to anticipate the turn of the century. Using 2 digits instead of 4 digits made data systems unable to distinguish the year 1900 from 2000. As the year '00' approached, press reports hyped up the prospect of worldwide doom. Most companies were able to fix the problem before the deadline, except a few places like the U.S. Naval Observatory, where the clock temporarily displayed the in-

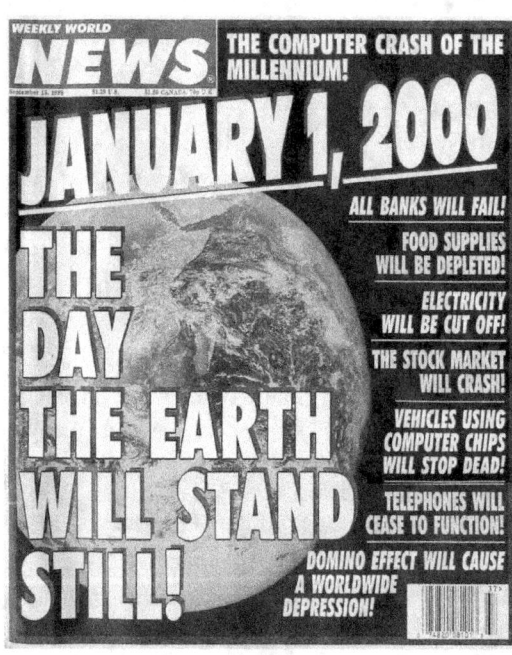

correct date. Worldwide, the cost of fixing the Y2K problem was $208 billion. The first Y2K-related lawsuit was from an upscale grocer. A 1997 credit card caused the crash of their 10 cash registers, repeatedly, due to year 2000 expiration dates. Some taxi meters stopped working, or incorrect taxi fares were given. 10,000 card swipe machines stopped processing credit and debit card transactions. On January 1, 2000, radiation-monitoring equipment failed in Japan. In Australia, bus ticket validation machines failed. In the US, 150 Delaware Lottery casino slot machines stopped working. At the US Naval Observatory, which runs the master clock that keeps the country's official time, gave the date as 1 Jan 1900. In France, the national weather forecasting service made the date on a webpage showing a map of a different day.

Oddities of Science
Oddities in Geology

Acasta Gneiss. The Acasta gneiss (pronounced nice) is a type of metamorphic rock on a remote island in Slave Craton in Northwest Territories, Canada. The rock was metamorphosed about 4.3 billion years old and is the oldest known intact crustal fragment on Earth. It was first described in 1989 and named for the nearby Acasta River. It is the oldest known exposed rock in the world. The age is based on the radiometric dating of its zircon crystals using mass spectrometry at the ETH labs in Zurich, Switzerland. The Acasta Gneiss is important in establishing the early history of the continental crust. [source: Orwig, "Archean Rocks in the Acasta Gneiss Complex," Earth & Space Science News, Dec 31, 2014]

Barringer Crater. Barringer Crater, also known as Meteor Crater or Canyon Diablo Crater, is a meteorite impact crater between Flagstaff and Winslow, Arizona. In the 19th century, it was thought that the crater was the result of a volcanic steam explosion. In 1902, geologist Daniel Barringer (1860-1929) had a theory that the crater was produced by the impact of a large iron-metallic meteorite. Barringer staked a mining claim to the land and received a land patent signed by Theodore Roosevelt. He created the "Standard Iron Company" at the site in order to mine the crater for the iron that he assumed was buried below its surface. After drilling as deep as 1,375 feet for two years, the company was unable to find the meteorite beneath the surface. The mining of the crater continued until 1929 without finding the 10-million-ton meteorite that Barringer assumed must be hidden. Barringer died of a heart attack on November 30, 1929 after reading reports that no iron was to be found. Barringer had spent $600,000 in mining the crater, nearly bankrupting him, with no profits

to show for it. It wasn't until 1960 that the impact theory was correct, and the site was hit by a meteorite. The site is not protected by the government since the crater is privately owned by the Barringer family. The diameter of the crater is about 4,000 feet. It is about 600 feet deep. The crater has a square outline instead of a circular outline due to regional cracks in the strata at the impact site and erosion. The crater was created about 50,000 years ago by a meteor about 160 feet in diameter. It stuck about 29,000 miles per hour. It hit with the force of 10 megatons. The meteorite was mostly vaporized upon impact. During the 1950s and 1970s, NASA astronauts trained in the crater to prepare for the Apollo missions to the moon. In 1956, Clair Patterson (1922-1995) used lead isotopic data from one of the meteorites near the rim of the crater to calculate the age of the meteorite and the age of the Earth. Using lead isotropic date from the meteorite, he calculated the age of the Earth of 4.55 billon years. Before this discovery was made, it was believed that the Earth was 3.3 billion years old. [source: Tobin, "Meteor Crater Arizona – World's Best Meteorite Impact Crater," meteorite.com, Oct 21, 2016]

Cal Orko. Cal Orko, near Sucre, Bolivia, is the home of the world's largest group of dinosaur prints. There are over 12,000 dinosaur tracks on the 328-foot vertical quarry limestone slab. Nine different species of dinosaurs have been documented with 462 distinct dinosaur tracks. The discoveries of the dinosaur tracks were first made in 1985. The land surrounding this ancient watering hole was itself shifted upwards by subsequent tectonic movements. During the time of the dinosaurs, Cal Orko was near a huge lake which boasted the continent's first flowers. The attracted the herbivore dinosaurs, followed by the carnivore dinosaurs. [source: Belcher, "Dinosaur tracking in Bolivia," *The Guardian*, July 15, 2011]

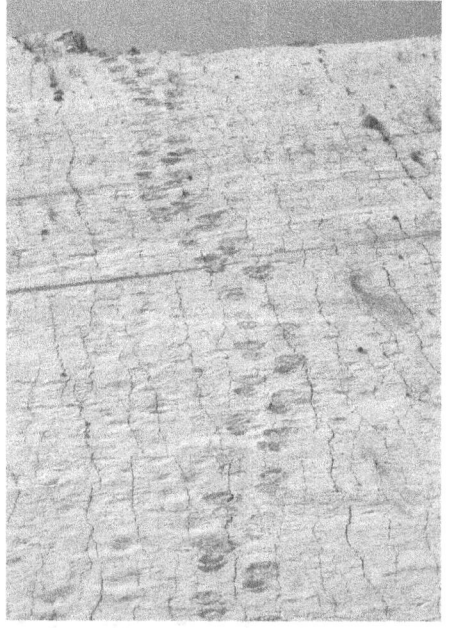

Clear Lake Volcanic Field. The Clear Lake Volcanic Field is a volcanic field beside Clear Lake 90 miles north of San Francisco, California. It is one of the oddest volcanoes in the world. Its formation and existence is still a mystery. It does not sit on a subduction zone, nor a rift zone, nor a hot spot. The field's magma chamber powers a geothermal field called The Geysers, which holds the largest complex of geothermal power plants in the world. Its magma chamber is estimated to be around 1,400 cubic kilometers, one of the largest in the world. Clear Lake does not have any known eruptive pattern. Eruptions were fairly active from 60,000 years ago until 10,000 years ago, and since then they have all but stopped.

Darvaza gas crater. The Darvaza gas crater, also known as the Door to Hell, is a natural gas field collapsed into a cavern in Derweze, Turkmenistan. In 1971, the site was identified by Soviet engineers as a large oil field site. A drilling rig was set up over a natural gas pocket, but the ground beneath the drilling rig and camp collapsed into a wide crater (about the size of a football field) and was buried. The engineers wanted to burn the gas off, so they lit it on fire and expected that the gas would burn out within a few weeks. It has now been burning for 49 years, and is expected to keep on burning. [source: Davies, "Door to Hell," *Science Alert*, Jan 31, 2017]

Devils Tower. Devils Tower, also known as the Bear Lodge Butte, is a rock formation formed by magma in the Black Hills of Wyoming that stands 867 feet from summit to base. It was the first declared United States National Monument, established in 1906 by President Theodore Roosevelt. In 1875, Colonel Richard Irving Dodge (1827-1895), during an expedition, named it Devils Tower when his Indian interpreter misinterpreted a native name to mean "Bad God's Tower." The  rock formation was featured in the movie *Close Encounters of the Third Kind*. [source: "Devils Tower," *National Geographic*, Jan 10, 2012]

Diamond. Diamond is a solid form of the element carbon. Diamonds are not formed from coal. Diamonds were formed over 1 billion years ago at high pressures over 100 miles below the surface in the Earth's mantle. Coal, known as fossil fuel, is formed from the dead remains of vegetation. Diamond has the highest hardness, least compressible, and highest thermal conductivity of any natural material. Sound travels faster in a diamond than any other substance. They were carried to the surface in volcanic eruptions. Popigai crater in Siberia may have the world's largest diamond deposit, estimated at trillions of carats, and formed by an asteroid impact 35 million years ago. The Argyle diamond mine in Australia is the largest producer of diamonds by weight in the world. In meteorites, about 3% of the carbon is in the form of very small diamonds (diameters of a few nanometers). Roughly 49% of diamonds originate from Central and Southern Africa. Until the late 19th century, diamonds were found only in a few riverbeds in India and the jungles of Brazil. At that

time, the entire world production of gem diamonds were about a few pounds a year. In 1870, huge diamond mines were discovered near the Orange River in South Africa. In response to economic conditions, diamonds have continued to advance upward in price every year since the Depression. In 1938, three-fourths of all diamonds were sold for engagement rings in the United States, thanks to good advertising that diamonds were "forever." Today, about 85% of rough diamond distribution is done by the De Beers Group, founded in 1888. The largest diamond in North American, 12.42 carats, was found in Arkansas in 1924. It is called the "Uncle Sam Diamond." [source: Epstein, "Have You Ever Tried to Sell a Diamond," The Atlantic, February 1982]

Kaffeklubben Island. Kaffeklubben Island, also called Coffee Club Island, is a small island lying off the northern tip of Greenland. It contains the northernmost point of land on Earth. In 1900, explorer Robert Peary (1856-1920) first recorded sighting of the island. In 1921, the island was first visited by the Danish explorer and geologist Lauge Loch (1892-1964). He set foot on the island and named it Kaffeklubben Island after the coffee club at the University of Copenhagen Geological Museum. The island is 443 miles from the North Pole. The island is about a half mile long and 980 feet across at its widest point. The highest point is 98 feet above sea level. Despite the harsh environment, flowering plants do grow on the island. [source: "The Flowers of Kaffeklubben Island – Greenland," *Atlas Obscura*]

Kali Gandaki Gorge. The Kali Gandaki (Andha Galhi) Gorge in Nepal is the deepest gorge in the world. On the west side of the gorge is Mount Dhaulagri at 27,795 feet. On the east side of the gorge is Annapurna at 26,545 feet. If one measures the depth of a canyon by the difference between the river height and the heights of the highest peaks on either side, the gorge is the world's deepest. The exact depth of the canyon remains unknown due to disagreement over rim height. The river in the gorge is 18,278 feet lower than Annapurna. As tectonic activity forced the mountains higher, the river cut down through the uplift. The gorge is older than the Himalayas of Nepal. The river source coincides with the Tibetan border and a Ganges tributary. The fossils of shellfish found in the area are called saligrams and are considered sacred stones.  They are valued more than gold in India because they are supposed to have unique healing properties. [source: Bensal, "The Kali Gandaki Gorge, Mustang, Nepal," *sites.miis.edu*, March 22, 2013]

Lake Baikal. Lake Baikal is located in southern Siberia. It is the largest freshwater lake by volume in the world. It holds 23% of the world fresh surface water. It exceeds a depth of 5,387 feet in places. Its lowest point lies more than 4,000 feet below sea level. This lake contains more water than all five of the Great Lakes combined and is also the deepest lake in the world. It is considered among the world's clearest lakes, and it is possible to see more than 130 feet down. It is also considered the oldest lake in the world, created 30 million years ago. It was formed as an ancient rift valley, where the Earth's crust is slowly pulling apart. The lake is fed by 330 inflowing rivers. [source: Szalay, "Lake Baikal: World's Largest, Deepest Lake," *Live Science*, Jan 27, 2017]

Macquarie Island. Macquarie Island is the only place on Earth where the Earth's mantle is above water. The island is midpoint between Antarctica and the island of Tasmania. The island is home to the entire royal penguin population during their annual nesting season. The island was discovered accidently in 1810 by the Australian sealer Frederick Hasselborough, who drowned 4 months after his discovery. He had been looking for new sealing grounds and found the island populated by as many as 400,000 seals. He claimed the island for Britain and named it after Lachian Macquarie (1762-1824), the Governor of New South Wales at the time. Macquarie Island is the result of the game geologic processes that created the Himalaya mountain range. It is located at the edge of two tectonic plates and experiences frequent earth tremors. The island is 1,345 feet at its highest point, exposing segments of oceanic crust and parts of the Earth's mantle above water. [source: "Macquarie Island station: a brief history," *antarctica.gov.au*, Aug 7, 2018}

Maldives. The Maldives is a small island nation located in the Arabian Sea of the Indian Ocean. It lies halfway between Indonesia and Africa. The Maldives archipelago is located on the Chagos-Laccadive Ridge, a vast submarine mountain range. The Maldives is made up of 26 atolls and 1,190 islands. The Maldives is one of the world's most geographically dispersed sovereign states. It is the smallest Asian country by land area and population. It's average ground-level elevation is less than 5 feet above sea level. It is the world's lowest-lying country. It highest natural point is 17 feet above sea level. The geology of the Maldives formed beginning 68 million years ago as a hotspot which produced the Deccan Traps in India, one of the largest volcanic features on Earth. [source: Schechter, "A Guide to Vacationing in the Maldives," *Travel and Leisure*, April 22, 2017]

McMurdo Dry Valleys. The McMurdo Dry Valleys sits in the Victoria Land region of Antarctica, to the west of McMurdo Sound. It is the world's driest desert and has not seen any precipitation in millions of years. It is also the coldest and windiest desert on planet Earth. It is completely dry and ice-free. Mountains surrounding it block coastal precipitation. Also, winds can blow as much as 200 miles per hour in the valley, heating the air and stripping out any moisture. No living organism has ever been found in the region. Scientists consider the Dry Valleys the closest thing to the environment of Mars and conduct tests there. 98% of the rest of Antarctica is covered in a mile-deep sheet of ice. [source: McMurdo Dry Valleys, *Atlas Obscura*]

Mount Thor. Mount Thor (Thor Peak), named for the Norse god of thunder, is located in the Baffin Mountains on Canada's Baffin Island. It s 5,495 feet in height, which features the world's greatest vertical drop – 4,101 feet, with the cliff overhanging at an average of 15 degrees from vertical. It owes its sheerness and steepness to the intense glaciations that scoured the Canadian arctic during the last ice age. In 1965, Donal Morton and Lyman Spitzer made the first ascent of the mountain. The peak draws a large number of rock climbers, base jumpers, and hang gliders. [source: Jennings, "Meet Canada's Mount

Oddities of Science

Thor: The World's Steepest, Tallest Cliff," *Conde Nest Traveler*, June 10, 2013]

Nazca Lines. The Nazca Lines are a group of very large designs 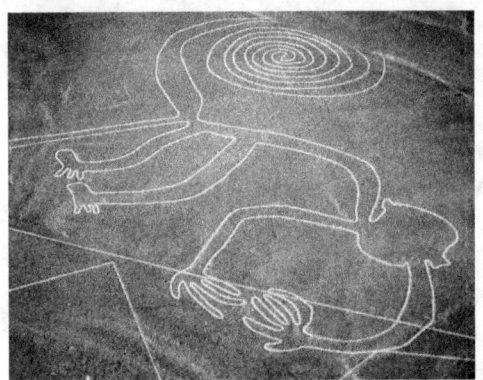 (geoglypths) in the ground located in the Nazca Desert in southern Peru (one of the driest deserts on Earth). They are formed by depressions made in the soil, with the lines 6 inches to a foot wide and 6 inches deep. They were formed over 1,700 years ago. Most lines run straight across the landscape, but some are figurative designs of animals and plants. The lines cover about 170 square miles. If you add up all the length of the lines, it would be over 800 miles. The largest continuous line is 1,200 feet long. Some of the lines form shapes that are best seen around 1,500 feet in the air. There are over 70 animal designs, along with trees and flowers. In 1553, the lines were discovered by Spanish conquistadors. They thought they were trail markers. In the 20th century, Peruvian military and civilian pilots observed and reported them. Since 2006, over 100 new geoglyphs have been found. Oddly, the most-depicted designs are exotic birds found only in non-desert habitats far from this desert. In 2019, researchers analyzing satellite and high-resolution imaging data have found 143 new Nazca lines. They include a humanoid figure 16 feet across that was spotted by IBM's Watson AI system (IBM Watson Machine Learning Community Edition).This represented the first glyph discovered by Artificial Intelligence (AI). [source: "143 New Geoglyphs Discovered on the Nazca Pampa and Surrounding Area,"*Yamagata University News*, Nov 15, 2019]

Pelée. Mount Pelée is a volcano at the northern end of Martinique in the Lesser Antilles. Its last eruption was in 1932. The eruption on May 8, 1902 destroyed the town of Saint-Pierre. The city was known as the "Paris of the Caribbean." 30,000 people died from the eruption in the space of a few minutes. It was the worst volcanic disaster of the 20th century. Only 2 people survived. One person survived because he was in a poorly ventilated, dungeon-like jail cell. The other survived who lived on the edge of the city and escaped with severe burns.

[source: Rosen, "Benchmarks: Mat 8, 1902: The deadly eruption of Mount Pelée,: *Earth Magazine*, May 8, 2015]

San Francisco earthquake. On April 18, 1906, a magnitude 7.9 earthquake struck the coast of Northern California and centered on the San Andreas fault, striking near San Francisco. The earthquake was felt from southern Oregon to Los Angeles, to central Nevada. The earthquake and fire destroyed over 80% of the buildings in the city. It killed as many as 6,000 people and left 300,000 homeless from a population of about 410,000. 20,000 victims were rescued by the U.S.S. Chicago, one of the largest sea evacuations in history. Some 500 looters were shot by police and the military. It was one of the worst and deadliest earthquakes in the history of the United States. It was the greatest loss of life from a natural disaster in California's history. [source: "Great San Francisco Earthquake," *National Geographic*, Dec 16, 2013]

Tsunami. The world's tallest tsunami, with a height of 1,720 feet, occurred in Lituya Bay, Alaska. On July 9, 1958, an 8.3 magnitude earthquake hit the area and loosened 40 million cubic yards of rock (90 million tons) above the northeastern shore of Lituya Bay. The mass of rock plunged 300 feet down into the waters of Gilbert Inlet. The impact force of the rock fall generated a local tsunami that crashed against the southwest shoreline of Gilbert Inlet. One boat with 2 persons on board vanished without a trace into the open sea, carried away by the tsunami. The impact was heard 50 miles away. [source: Bressan, "World's Tallest Tsunami Hit the Gulf of Alaska More Than 60 Years Ago," *Forbes*, July 6, 2019]

Yarrabubba crater. The Yarrabubba crater in Western Australia is the oldest known impact structure on Earth. Its age has been calculated to be over 2.2 billion years old. The crater is 43 miles wide. Researchers pinpointed the crater's age by dating microstructures in crystallized rock (shocked quartz) that formed when the impact occurred. The rim of the original crater has been completely eroded, and is not readily visible on aerial or satellite images. A computer simulation suggests that a Yarrabubba-sized impact would have released 200 trillion kilograms of water vapor into the atmosphere, which could have warmed the planet and melted ice sheets. [source: Erickson, "Precise radiometric age established Yarrabubba, Western Australia, as Earth's oldest recognized meteorite impact structure, *"Nature Communications*, Jan 21, 2020]

Oddities of Science
Oddities in Medical Science

Bone. Babies are born with over 270 bones. Some fuse together as their bodies grow. An adult human skeleton has 212 bones. The smallest bone is the stapes in the middle ear. The femur (thighbone) is the longest and strongest.

Bone Marrow Transplant. A bone marrow transplant is a medical procedure performed to replace bone marrow (which produces blood cells) that has been damaged or destroyed by disease, infection, or chemotherapy. In 1968, the first bone marrow transplant took place at the University of Minnesota on a 22-month-old boy with severe combined immunodeficiency, more commonly known as "bubble boy disease." Stem cells are removed from a large bone of the donor, typically the pelvis, through a large needle that reaches the center of the bone. The technique is referred to as a bone marrow harvest. The boy's sister had donated the bone marrow, which allowed her brother to live into adulthood. [source: "First Successful Bone Marrow Transplant Led to Today's Immunotherapies,"*med.wisc.edu*, Jul 18, 2018]

Coca-Cola. Coca-Cola was first created as a medicinal tonic by biochemist and pharmacist John Pemberton (1831-1888) in 1866. It was alcoholic and also contained cocaine (9 milligrams per glass) as a major ingredient. It was created as an alternative to morphine. This was due to Pemberton, a Confederate Colonel, being addicted to morphine after he suffered a saber wound during a Civil War battle. He was looking for an alternative to the pain killer that was opium-free. The original name was Pemberton's French Wine Coca and he sold it as a "brain tonic and intellectual beverage." The two primary ingredients were the coca leaf and the kola nut. Extract of the coca leaf provided the cocaine, and the kola nut provided caffeine. In 1886, the temperance movement forced Pemberton to produce a non-alcoholic alternative to his French Wine Coca. By accident, he mixed syrup with carbonated water and decided to sell this as a fountain drink rather than medicine. It still contained cocaine. Over the years, as the recipe was refined, the amount of cocaine was reduced bit-by-bit. By 1929, the beverage was cocaine-free. [source: Myers, "25 Things You Didn't Know About Coca-Cola," *The Daily Meal*, June 13, 2019]

Coronavirus. Coronaviruses are a group of related viruses that cause diseases in mammals and birds. Coronavirus disease 2019 (COVID-19) is an infectious disease caused by a newly discovered coronavirus. It is caused by the severe acute respiratory syndrome (SARS) coronavirus 2 (SARS-CoV-2). 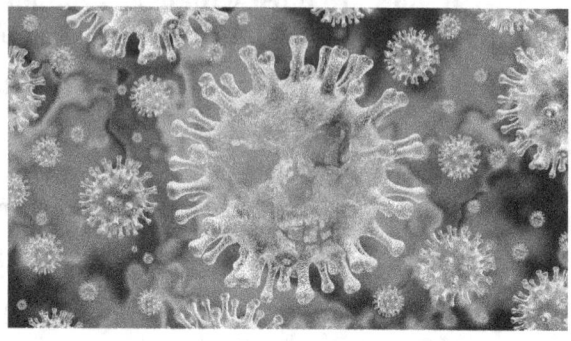 The disease has spread globally since December 2019, resulting in a pandemic. The fatality rate is between 1% and 8%. For those over 80 years of age, the fatality rate is 10% to 27%. The virus is thought to have an animal origin. It was first transmitted to humans in Wuhan, China in November or December 2019, perhaps by bat droppings to another animal, before jumping to humans. Earlier coronaviruses were discovered in the 1960s, found in chickens. The name "coronavirus" is derived from the Latin corona, meaning "crown" or "wreath." The name refers to the characteristic appearance of the virions (the infective form of the virus) by electron microscopy. A halo of spiked proteins adorn the outer surface, which resemble a crown (corona) under and electron microscope. The virus invades and replicate in the lung cells. A buildup of debris and fluids occurs, causing shortness of breath, and, in some cases, pneumonia. This new virus had not infected people before the outbreak in China, so no one had any prior immunity to it. The coronavirus can live in air for 3 hours and on surfaces for 3 days. One sneeze or cough can contain 6 million coronaviruses. [source: "Coronavirus Resource Center, *Johns Hopkins University & Medicine*]

CRISPR. CRIPR stands for Clustered Regularly Interspaced Short Palindromic Repeats. It is a gene-editing tool that allows scientists to modify DNA and genes. It is a family of DNA repeats found in bacteria and archaea (single-celled organisms). They are used in medicine and other disciplines to detect and destroy DNA in infections. The CRISPR gene-editing system uses an enzyme called Cas9 and a customized guide RNA to help target, cut, alter, and stitch up particular stretches of the genome. It can disrupt a gene or edit a new one. The technology can be used to edit genes within organisms. The editing process can be used in treatment of diseases and the development of biotechnology products. In 1987, a Japanese researcher accidently cloned part of a CRISPR sequence together with a certain gene. He discovered clustered DNA repeats. In 2001, Francisco Mojica and Ruad Jansen proposed the acronym CRISPR to describe this genetic oddity. Its technology has been applied in the food and farming industries to immunize cultures such as yogurt and to increase yields in crops, making them drought tolerant. In medicine, they have been used to inactivate genes in human cells. CRISPR has also been used to change mosquitoes so they cannot transmit diseases such as malaria. In 2019, CRISPR was used by doctors to successfully treat a patient with sickle cell disease. Currently, there are clinical trials using CRISPR to treat cancers such as non-Hodgkin's lymphoma. A new CRISPR genome-editing tool uses enzymes that can precisely target DNA without introducing unwanted mutations. This allows researchers to convert one DNA letter into another. This can be used to treat genetic diseases such as sickle cell anemia, which is caused by a single-letter change in the DNA.

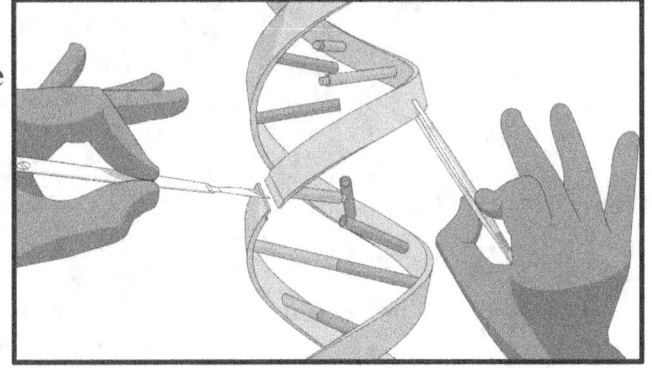

[sources: Super-precise CRISPR tool enhanced by enzyme engineering, *Nature*, Feb 10, 2020 and "Editing the Genome," *Scientific American*, January 2020, p. 23]

Flu pandemic. A flu (influenza) pandemic is an epidemic of the flu virus that spreads worldwide and infects are a large proportion of the world population. There have been 9 flu pandemics during the last 300 years, with the most recent being in 1918, 1947, 1957, 1968, 1977, 1989, and 2009. The 1918 influenza pandemic (Spanish flu) lasted from January 1918 to December 1920. The Spanish flu infected 500 million people (1/3 of the global population) and killed far more people than all of World War I that were killed or wounded. About 50 to 100 million people died from the Spanish flu, or about 3% of the world's population. During World War I, 20 million died and 21 million were wounded. In the USA, 675,000 Americans died of the Spanish flu. Although it gained the nickname "Spanish flu," it is unlikely that the virus originated in Spain. The first case was a man in Kansas. At one point, the use of aspirin was blamed for causing the pandemic, when it might actually have helped those infected. In New York City, it was illegal to spit on the streets during the pandemic. Stores could not hold sales and funerals were limited to 15 minutes. In 1919, President Woodrow Wilson (1856-1924) contracted the disease while negotiating the treaty of Versailles. [source: "Spanish flu: the deadliest pandemic in history," *Live Science*, March 12, 2020]

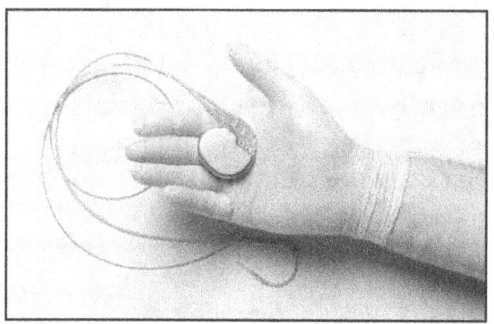

Heart pacemaker. The first heart pacemaker ("artificial pacemaker") was designed in 1926 by Dr. Mark Lidwill (1878-1969 and Edgar Booth, The first external pacemaker plugged into a wall socket. Lidwill was born in England but settled in Australia. His first pacemaker helped resuscitate a newborn baby with an electrical device that administered 16-volt impulses, 80 to 120 pulses per minute, to the heart via a needle put into the heart. In 1950, John Hopps of Canada designed and built the first external pacemaker. It was powered from an AC wall socket which had the potential of electrocuting the patient. In 1952, the pacemaker was powered by a large rechargeable battery as the power supply. Today, a pacemaker's battery lasts 5 to 15 years. Today's pacemakers adjust the heart rated to match the level of physical activity. [source: "Pacemaker," *Mayo Clinic*, June 25, 2018]

Hospital. The first official medical department or "hospital" in America was started by the Army in New York following the Battle of Bunker Hill in 1775. It consisted of 6 surgeons, 20 surgeons' mates, a nurse for every 10 men, a clerk, and 2 storekeepers. A Revolutionary War soldier faced a 2% chance of dying in battle, and a 25% chance dying in an overcrowded army hospital. Six out of seven deaths during the Revolutionary War were due to camp illness. For every one death from wounds, there were nine deaths from disease. Life expectancy in the army was 35 years. In all of America, there were only 300 to 400 medical doctors with medical degrees. There were earlier unofficial hospitals in America, but they were hastily built to confine contagious diseases during epidemics, primarily in seaport towns.

Kidney transplant. The main function of the kidneys is to filter and remove waste, minerals, and fluid from the blood by producing urine. In 1933, the first attempt of a human kidney transplant was made by Yuriy Vorony (1895-1961). He used a kidney removed 6 hours earlier from a deceased donor and reimplanted it in the patient's thigh. The patient died 2 days later, as the graft was incompatible with the patient's blood group. In 1950, the first successful transplant was performed by Dr. Richard Lawler (1896-1982) in Illinois. The patient lived another 5 years. In 1954, Dr. Joseph Murray (1919-2012) performed the first successful human kidney transplant on identical twins. Murray shared the Nobel Prize in Medicine with Donnall Thomas for their discoveries concerning organ transplants. [source: "Kidney transplant," *Mayo Clinic*, Feb 25, 2020]

Lobotomy. A lobotomy, also known as a leucotomy, is a form of neurosurgery that involves severing connections in the brain's cortex. In the 20th century, lobotomies were legitimate treatments for various mental health illnesses, despite general recognition of frequent and serious side effects that took away a person's personality and intellect. In 1935, Portuguese neurologist Antonio Egas Monitz (1874-1955) originated the procedure in Lisbon on the brains of the mentally ill. Moritz actually did not perform the surgery himself. He was not trained in neurosurgery and his hands were crippled from gout. The procedure, by drilling through the skull, was performed by his assistant, Pedro Lima. After the surgery, Moritz declared her "cured" although she was never released from the mental hospital. Monitz shared the Nobel Prize for Medicine in 1949 for his work. In 1946, Walter Freeman performed the first "ice-pick" 10-minute lobotomy in the United States and gave "lobotomy" its name. By 1951, there were almost 20,000 lobotomies that had been performed in the United States. In the United States, about 40,000 people were lobotomized between 1936 and 1970. The

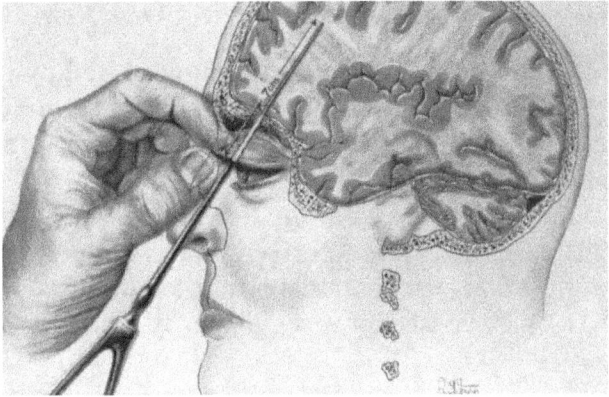

Soviet Union banned the practice in 1950 on moral grounds. The majority of lobotomies were performed on women. By the late 1970s, the practice of lobotomy had generally ceased, although it continued as late as the 1980s in France. [source: 'My Lobotomy': Howard Dully's Journey, *NPR*, Nov 16, 2005]

Medicinal Plants. Medicinal plants have been discovered and used in traditional medicine practices since prehistoric times. The oldest written evidence of medicinal plants' usage for preparation of drugs has been found on a Sumerian clay slab that is about 5,000 years old. 25% of all prescribed medicines are derived from just 40 plants. Of the top 150 prescription drugs, 118 are based on natural sources. In 1806, morphine was discovered from the poppy flowing plant. In 1820, quinine was first isolated from the bark of a cinchona tree. Anti-cancer drugs were developed from the yew and the Madagascar periwinkle. Over 50,000 medicinal plants are used across the world. Scientists have documented over 18,000 plant species (out of a total of 391,000 plant species) that have a medicinal use. Drugs derived from plants include opiates (opium and codeine), cocaine, and marijuana. There has been an accelerating loss of plant species worldwide. It is estimated that 15,000 medicinal plant species are threatened with extinction. [source: Deering, "Nature's 9 Most Powerful Medicinal Plants and the Science Behind Them," *Healthline*, Feb 28, 2019]

Plastic Surgery. Plastic surgery is a surgical specialty involving the restoration, reconstruction, or alteration of the human body. The word plastic comes from the Greek word "plastikos" which means "to shape or mold." The surgical definition of "plastic" first appeared in 1938. The most popular cosmetic surgeries, in order, are breast augmentation, liposuction, eyelid surgery, nose job, and tummy tuck. [source: "Cosmetic Plastic Surgery Overview," *Johns Hopkins Medicine*]

Polio. More polio cases are now caused by the polio vaccine than by the wild virus. Recently, four African countries have reported new cases of polio linked to the oral vaccine. Global health numbers now show there are more children paralyzed by viruses originating in vaccines than in the wild. In 2019, there were 125 cases of wild polio and 241 cases of vaccine-derived polio. Polio occurs naturally only in humans and is preventable with the polio vaccine. [source: "More polio cases now caused by vaccine than by wild virus." *AP News*, Nov 25, 2019]

Oddities of Science

Smallpox. Smallpox is an infectious disease caused by one of two virus variants. The last major outbreak of smallpox was in 1972 in Yugoslavia. The last case of smallpox was diagnosed in 1977. Routine childhood vaccination of smallpox was discontinued in the United States in 1972. The origin of smallpox is unknown, probably from rodents. The earliest evidence of smallpox dates to the 3rd century BCE in Egyptian mummies. Smallpox was found in the study of the Egyptian mummy of Ramses V (died in 1145 BC). The epidemic of 735-737 killed one-third of the population of Japan. Fatality rates of smallpox during outbreaks in Native American populations in the 1600s killed 80-90% of those infected by smallpox. Over 400,000 people died per year of smallpox in the 18th century. Smallpox was the principal cause of death in Aboriginal population in Australia between 1780 and 1870. 300 million died of smallpox in the 20th century. 500 million people have died of smallpox since 1870. In 1967, there were over 15 million cases of smallpox that year.  Smallpox is only one of two infectious diseases that have been eradicated. The other disease that has been eradicated is rinderpest (a cattle disease), last seen in 2011. In 1796, Dr. Edward Jenner, a physician in rural England, discovered that immunity to smallpox could be produced by inoculating a person with material from a cowpox lesion. Jenner noticed that dairy farmers did not get smallpox. Jenner called the material used for inoculation vaccine, from the root word vacca, meaning cow in Latin. In 1796, the first child to get vaccinated in Russia was given the name "Vaccinov" by Catherine the Great. [source: Bradford, "Smallpox: The World's First Eradicated Disease," *Live Science,* Apr 23, 2019]

Thalidomide. Thalidomide, sold under the brand name Thalomid, was first marketed in West Germany in 1957. When first released, it was promoted for anxiety, trouble sleeping, tension, and morning sickness. At first, it was deemed to be safe in pregnancy, but the medicine was not fully tested. By 1961, noticeable birth defects in the thousands were occurring after mothers started taking the drug. The total number of people affected by use during pregnancy was estimated at

10,000, of which about 40% died around the time of birth. Those who survived had limb, eye, urinary tract, and heart problems. The thalidomide tragedy prompted the United States and international regulatory agencies to develop systematic toxicity testing procedures. [source: Kim & Scialli, "Thalidomide: the tragedy of birth defects and the effective treatment of disease," *Toxicological Sciences*, Feb 2012]

Vampire Syndrome. Humans need sunlight to synthesize vitamin D, but too much exposure to the Sun's ultraviolet (UV) rays can damage the skin. About one in a million people have what is known as xeroderma pigmentosum, which is commonly known as XP, and are extremely sensitive to UV rays. These people must be completely shielded from sunlight, or will experience extreme sunburns and breakdown of the skin. They could easily develop skin cancer. Xeroderma pigmentosum, also called the Vampire Syndrome, is caused by a rare recessive mutation of one of the repair enzymes. Functioning normally, these enzymes correct any damaged DNA that can be caused by UV rays. For those with this condition, the enzymes do not work properly, and the DNA gets further damaged. The best prevention is to stay completely out of sunlight, just like a vampire. Another disease, called porphyria, also called the vampire's disease, has the symptoms of extreme sensitivity to sunlight and the discoloration of the urine.

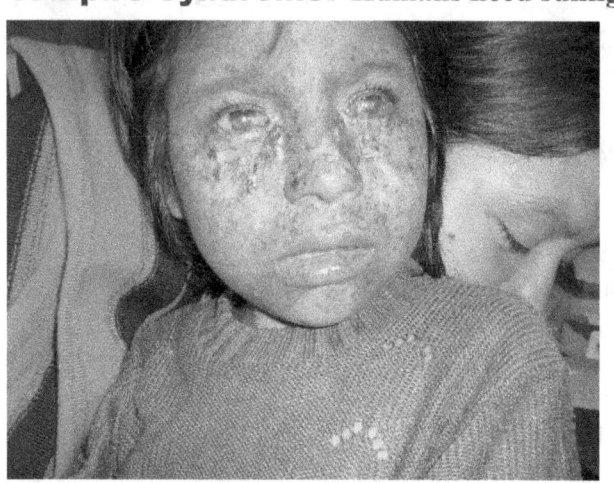

Oddities of Science

Oddities in Meteorology

Blizzard. A blizzard is a severe snowstorm characterized by strong sustained winds and lasting for at least three hours. The difference between a blizzard and a snowstorm is the strength of the wind, not the amount of snow. To be a blizzard, a snowstorm must have sustained winds or frequent gusts greater than 35 miles per hour. In Antarctica, blizzards are associated with winds that blow 100 miles per hour or greater. The Great Blizzard of 1888 dropped 50 inches of snow with winds of more than 45 mph. Snow drifts were over 50 feet high. It killed 400 people, mostly in New York. In 1972, a blizzard in Iran killed 4,000 people. It dumped 26 feet of snow, burying 200 villages. It was the deadliest blizzard in recorded history. After a snowfall lasting nearly a week, an area the size of Wisconsin was entirely buried in snow. In March 1993, the Storm of the Century wreaked havoc from Cuba to Canada. It was both a blizzard and a cyclone. It was as strong as a hurricane, covering the entire continent. It killed 310 people and caused $6.6 billion in damage. It shut down the South for 3 days. [Source: Oskan, "The 10 Worst Blizzard in US History," *Live Science*, Feb 8, 2013]

Cirrus cloud. Cirrus clouds are characterized by thin, wispy stands. They form at any altitude between 16,500 feet and 45,000 feet above sea level. They form so high in the sky that only ice crystals can form up there. The atmospheric temperature is -4 to -22 degrees Fahrenheit. New research finds that the cirrus clouds condense and freeze on very specific mineral and metal particles high in the atmosphere. Cirrus clouds cover up to 30% of the Earth's atmosphere at any given time. [source: Main, "How Cirrus Clouds Form – And Why It Matters," *Live Science,* May 9, 2013]

Cyclone Tracy. Cyclone Tracy was a tropical cyclone that devastated Darwin, Northern Territory on Christmas Day, December 25, 1974. Wind gusts were reported in the city at about 135 mph before instruments failed. Later estimates show wind speeds of 150 mph. It left 9 out of 10 homeless. Residents of Darwin were celebrating Christmas, and did not immediately acknowledge the emergency. The city was alerted to an earlier cyclone (Selma) that passed west of the city. Additionally, news outlets had only a skeleton crew on duty over the holiday. Cyclone Tracy killed 71 people and destroyed 80% of houses in Darwin. Over 30,000 people of the 40,000 residents were later evacuated, of who many never returned. The storm also sunk a Royal Australian Navy patrol boat, killing 2 sailors. Cyclone Tracy was Australia's worst ever Cyclone that caused $837 million ($7 billion in today's currency) in damage. [source: McLennan, "Tropical Cyclone Tracy was the worst of them all," *The Courier* (Australia), March 27, 2017]

Fire tornado. A fire tornado is a phenomenon that often occurs during large fires and wildfires. They resemble tornadoes in shape

and form in a similar was as a regular tornado. However, a fire tornado's vortex of the fire whirl doesn't extend from the ground to the base of the clouds. They get their shape from the heat of the fire itself mixed with surface winds. Temperatures inside a fire tornado can reach as high as 2,000 degrees Fahrenheit. They can reach up to 400 feet in height and 50 feet wide, with a wind speed up to 300 miles per hour. In 2018, during the Carr Fire in Shasta County, California, a fire tornado had a diameter of 1,000 feet and rose to 40,000 feet. The Carr fire tornado lasted over 1.5 hours. It was the strongest fire-induced tornado-like circulation ever recorded. [source: Rice, "California 'fire tornado' had 143 mph winds, possibly state's strongest twister ever," *USA Today*, Aug 3, 2018]

Grayson Storm. On January 3-4, 2018, Winter Storm Grayson hit the Southeast. It was one of the most intense winter storms in decades. It was one of the heaviest one-day snowfalls on record in Georgia and South Carolina (7.3 inches near Charleston). This was the first time since January 1885 that snow fell in the Florida panhan-

Oddities of Science

dle in the month of January (snowflakes appeared in February 1989). Over 3 inches of snow were measured in Tallahassee. Wind gusts were reported in excess of 70 mph in the Outer Banks of North Carolina. Top winds of 77 mph were measured on Fisherman's Island, Virginia. It brought snow all the way up to Maine, where it dropped 22 inches of snow in Etna and Plymouth, Maine. [source: Winter Storm Grayson Southeast Recap: Snow and Ice From North Florida to the Eastern Carolinas, *The Weather Channel*, Jan 4, 2018]

Hail. Hail is a form of solid precipitation, consisting of balls or irregular lumps of ice, called hailstones. Hail is formed when drops of water freeze together in the cold upper regions of thunderstorm clouds. Hailstones are not frozen raindrops. Hailstones can fall as fast as 110 mph. The heaviest hailstone fell in Bangladesh in 1986, weighing 2.25 pounds. That hailstorm killed 92 people. The largest hailstone fell in South Dakota in 2010, about 8 inches in diameter (about the size of volleyball). Hail can do a lot of damage. Hail is one of Canada's most expensive hazards. In the 9th century, hail killed about 600 nomads in India. In 1360, a hailstorm outside Paris killed hundreds of invading English soldiers. Because of this, King Edward II soon gave up his conquest of France. In 1888, hail killed more than 250 people in Moradabad, India. In 2014, up to 4 feet of hail fell on one city block in Denver, Colorado. [source: Pappas, "How Big Was the Biggest Hailstone Ever?" *Live Science,* Jul 12, 2010]

Hurricane. The strongest and most intense hurricane on record was Hurricane Patricia in October 2015. This Category 5 hurricane had a maximum sustained (over one minute) winds of 215 miles per hour off the southwest Mexican coast. In a 24-hour period, Patricia's maximum sustained winds increased by 120 mph, the fastest such intensification of a storm ever recorded. Its lowest central pressure was 872 millibars, only 2 millibars from the global record set by Super Typhoon Tip in 1979. Hurricane Patricia made landfall north of Manzanillo, Mexico, a sparsely populated area. Only two deaths were directly attributed to Hurricane Patricia, from a fallen tree on a campsite. Hurricane Patricia set

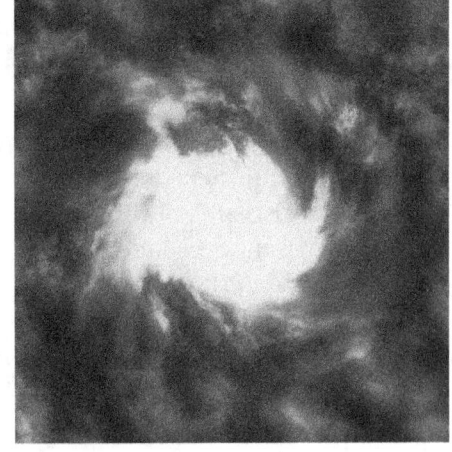

multiple records for maximum strength, rate of intensification, and rate of weakening throughout its short existence. In the United States, there are, on average, 1,200 tornadoes every year. In 1950, hurricanes were named alphabetically with the radio alphabet. By 1953, the National Hurricane Center was using alphabetical lists of women's names to name hurricanes. [source: Belles, "Three Years Ago, Hurricane Patricia Became the Record Strongest Hurricane in the Western Hemisphere with 215 MPH Winds," *weather.com*, Oct 22, 2018]

Lightning. The most persistent and most spectacular thunderstorm in the world is called the Catatumbo Lightning in Venezuela. It occurs 140 to 160 nights a year over the mouth of the Catatumbo River. This is a nearly continuous thunderstorm with up to 20,000 flashes of lightning per night. The lightning storm can last up to 10 hours long with 300 lightning strikes per hour. It produces about 10% of the ozone in the world. About 2,000 thunderstorms raid down on Earth every minute. [source: Schrader, "Venezuela's Neverending Thunderstorm," *TripSavvy*, June 3, 2019]

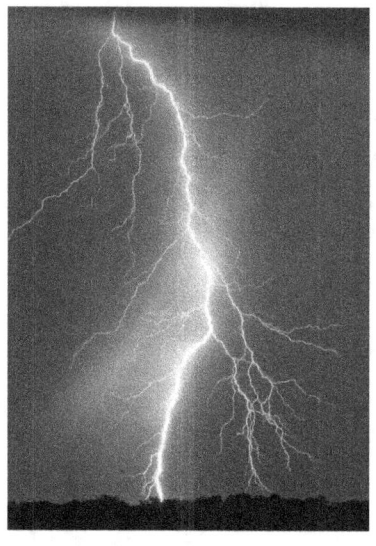

Ol Doinyo Lengai. Ol Doinyo Lengai is an odd active volcano in Tanzania, Africa. It is the only place on the planet that is currently erupting natrocarbonatite lava. Whereas most lavas are rich in silicate minerals, this volcano erupts lava made of carbonatite, which is rich in the rare sodium and potassium carbonates, nyereseite and gregoryite. Most volcanoes erupt at 1200 degrees , but Ol Doinyo Lengai erupts at a relatively low temperature of about 500 degrees Celsius (950 degrees Fahrenheit). This temperature is so low that the molten lava appears black, rather than having the red glow common to most lavas. After cooling, the lava appears gray or white. The resulting volcanic landscape is different from any other in the world. [source: Klemetti, "Strangest Magma on Earth: Carbonatites of Ol Doinyo Langai," *wired.com*, Mar 11, 2014]

Raindrop. Most of us imagine that raindrops look like tears, or the stretched out drip from a leaky faucet. Raindrops are actually shaped like the top of a hamburger bun, round on the top and flat at the bot-

Oddities of Science

tom. Because the airflow on the bottom of the raindrop is greater than the airflow on the top of the raindrop, this creates pressure on the raindrop's bottom, and its shape becomes flattened, like a hamburger bun, or punched in, so it looks like a kidney bean. The fastest speed a falling raindrop can hit you is 18 miles per hour. In 1995, the largest raindrops ever recorded were observed by meteorologists in Brazil. The size was 1 centimeter (.4 inches). The average lifetime of a raindrop is 20 minutes. On June 15-16, 1995, Cherrapunji, India endured 98.15 inches of rain, the most rain in any 48-hour period.

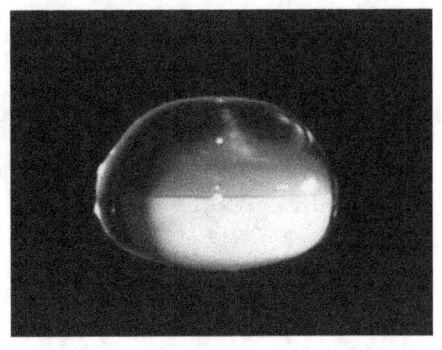

Temperature. Successful meteorology requires precise measurement of temperatures. Here are some of the extreme temperatures measured. The coldest temperature ever measured on Earth was -129 Fahrenheit (-89.2 Celsius) at the Soviet Vostok Station in Antarctica on July 21, 1983. In August 2010, satellite observations showed a surface temperature of -135.8 Fahrenheit (-93.2 Celsius) on dome Fuji in the eastern part of Queen Maud in Antarctica. In the United States, the coldest temperature recorded was in Rogers Pass, Montana at -69.7 Fahrenheit (-56.5 Celsius) on January 20, 1954. The highest temperature ever recorded and endorsed by the World Meteorologist Organization, was 134 Fahrenheit (56.7 Celsius) in Greenland Ranch, Death Valley desert, California on July 10, 1913, but this has been challenged. The temperatures reported in Death Valley at the time were not consistent with weather observed in the same region at the same time. The weather observer at the time was inexperienced and may have fudged the numbers. If the 1913 measurements are invalid, then the highest temperature ever recorded was

129 Fahrenheit (54 Celsius) in the Death Valley desert on June 20, 2013. The same temperature was also recorded in Kuwait on July 21, 2016. The highest natural ground surface temperature was 201 Fahrenheit (93.9 Celsius) at Furnace Creek, Death Valley, California on July 15, 1972. [source: Samenow, "New analysis shreds claim that Death Valley recorded Earth's highest temperature in 1913," *Washington Post*, Oct 25, 2016]

TIROS. On April 1, 1960, the first successful weather satellite, TIROS-1, was launched by NASA from Cape Canaveral. TIROS stood for Television Infra-Red Observation Satellite. Weather forecasting was deemed the most promising application of space-based observations. TIROS provided the first accurate weather forecasts based on data gathered from space. The spacecraft weighed 270 pounds and was equipped with two cameras and two video recorders. Image resolution was 1,000 feet on Earth. TIROS orbited 450 miles above Earth. The satellite operated for 78 days and took over 19,000 pictures. [source: "Celebrating the World's First Meteorological Satellite: TIROS-1," nesdis.noaa.gov, April 1, 2016]

Oddities of Science
Oddities in Physics

Big Bang. When Albert Einstein (1879-1955) developed the Theory of General Relativity, it predicted a possible expansion of the universe. In 1927, astronomer, physicist and priest Georges Lemaitre (1894-1966) thought that this indicated the universe began in a single point. The name "Big Bang" was given by Fred Hoyle (1915-2001) while mocking the theory during a radio broadcast. Hoyle coined the term "Big Bang" on BBC radio's *Third Programme* broadcast in 1949. Hoyle found the idea that the universe had a beginning to be pseudo-science and that the universe was in a steady state as it always has been and always will be. In 1964, the cosmic microwave background (CMB) radiation was discovered and supported that the universe was began as a single point about 14 billion years ago and was expanding. The Big Bang (there is really no sound in space) is not an explosion of matter moving outward to fill an empty universe. Instead, space itself expands with time everywhere and increased the physical distances between comoving points. Some physicists speculate a  new kind of phase transition during the big bang. The energy that drove the accelerating expansion of the universe during the so-called inflation period, converted into heat energy that created the hot bath of plasma that evolved into our familiar universe. [source: Lincoln, "A New Idea Might Help Scientists Understand the Big Bang Better," *Forbes*, Nov 12, 2019]

Castle Bravo. Castle Bravo was the first in a series of high-yield thermonuclear weapon design tests that occurred at the Bikini Atoll in the Marshall Islands in 1954. On March 1, 1954, the device was the most powerful nuclear device ever detonated by the United States. It was the first lithium deuteride fueled thermonuclear weapon. The blast was 15 megatons, which was 2.5 more times powerful than the predicted 6 megatons, due to unforeseen additional reactions that took place. This led to the unexpected radioactive contamination to the east of Bikini Atoll. Even after 65 years, today's radiation levels at some locations are higher than in areas affected by the Chernobyl and Fukushima nuclear disasters. Castle Bravo was one of 67 U.S. nu-

clear tests conducted in the Marshall Islands from 1946 to 1958. In 2019, studies found that in several locations in and near the testing sites, ambient radiation levels as well as radioactivity levels in ocean sediments and food remain higher that maximum exposure limits. [source: Cortier, "Marshall Islands Nuclear Contamination Still Dangerously High," *Earth & Space Science News*, July 16, 2019]

Cosmic Rays. In 1910, Theodor Wulf (1868-1946) compared the radiation at the bottom and the top of the Eiffel Tower. He found that energy was coming from outside the Earth's atmosphere and it was being detected by his device. This radiation was from cosmic rays, but his results were not initially accepted. Cosmic rays were discovered in 1912 by physicist Dr. Victor Hess (1883-1964) using a balloon, which earned him the Nobel Prize in 1936. "Cosmic rays" was named by Andrews Millikan (1868-1953) in 1925 when he proved that the cosmic rays were, indeed, of extraterrestrial origin. They are high-energy protons and atomic nuclei which move through space at nearly the speed of light. Many things about cosmic rays remain a mystery more than a century later. Scientists are not sure where they are coming from. Most scientists suspect they come from supernovas (big star explosions), but oddly, cosmic rays appear to be uniform to observatories examining the entire sky. The term "ray" in cosmic ray is a misnomer, since it is not a ray. It was first thought that cosmic rays were thought to be mostly electromagnetic radiation, but that turns out to be wrong. Cosmic rays have about 40 million times the energy of particles accelerated by the Large Hadron Collider. In 2017, cosmic rays were used to discover a possible void in the Great Pyramid of Giza. The scientists found this cavity using muon tomography, which examines cosmic rays and their penetrations through solid objects. If the human eye could see cosmic rays, the moon would be brighter than the sun. [source: Collins, "The moon would look brighter than the SUN if the human eye could see cosmic rays: Incredible NASA images taken with a gamma ray telescope transform it into a fiery ball," *Daily Mail* (UK), Aug 16, 2019]

Egg. If you take a hard-boiled egg and rapidly spin it, it will stand up on one end. It is such an odd and strange phenomenon that physicists did not solve how it did this until 2002. The egg acts like a tippe top that when spun, it flips upside down when spun. As it turns out, what is actually happening is that the egg is spinning on not one, but two axis. Objects usually spin around their center of mass. But due to the egg's shape, its center of mass and the point the egg starts spinning around, are not aligned. The difference between the center of mass and the mid-point of the egg causes the egg to wobble. This, in turn, makes the egg start to rise, tilting the angle of the short axis as the egg is spinning around. You can use physics to distinguish between hard-boiled eggs and raw eggs by spinning them. Hard-boiled eggs are much easier to stop spinning because they are solid. And when spun, they spin at a very fast pace. The liquid inside a raw egg prevents it from stopping and provided enough drag to start a raw egg spinning again. In a raw egg, its center of gravity changes as the fluid inside the egg moves around. This results in a wobbling motion in order to achieve balance. It also does not spin as fast.

Large Hadron Collider. The Large Hadron Collider (LHC) is the world's largest and most powerful particle accelerator. It lies in a tunnel 17 miles in circumference and as deep as 574 feet below ground between the France-Switzerland border near Geneva. It first started up on September 10, 2008. The LHC is colder than outer space. It is kept at 1.9 Kelvin (-271.3 Celsius or -456 Fahrenheit), almost absolute zero. A cryogenic cooling system keeps it frigid for the sake of the 9,300 superconductor electromagnets, which sends 200,000 billion protons towards one another at a rate of 11,245 times a second. The LHC is capable of accelerating particles up to 99.99 percent the speed of light and smashing them together. At $9 billion, it is one of the most expensive scientific instruments ever built. [source: Mann, "What is the Large Hadron Collider?" *Live Science*, Jan 29, 2019]

Muon. The muon (from the Greek letter mu) is an elementary particle similar to the electron, but are 200 times the mass of an electron. It is classified as a lepton and not composed of any simpler particles. It is an unstable subatomic particle with a mean lifetime of 2.2 microseconds. They are made when cosmic rays slam into atoms in Earth's atmosphere. Every square foot of the planet is hit by roughly 1 muon per second. Muons have been used to penetrate the innards of pyramids and volcanoes, and spot missing nuclear waste. In 2017, muons helped archaeologists find a previously unknown chamber in Egypt's Great Pyramid. [source: Gibney, "Muons: the little-known particles helping to probe the impenetrable," *Nature*, May 24, 2018]

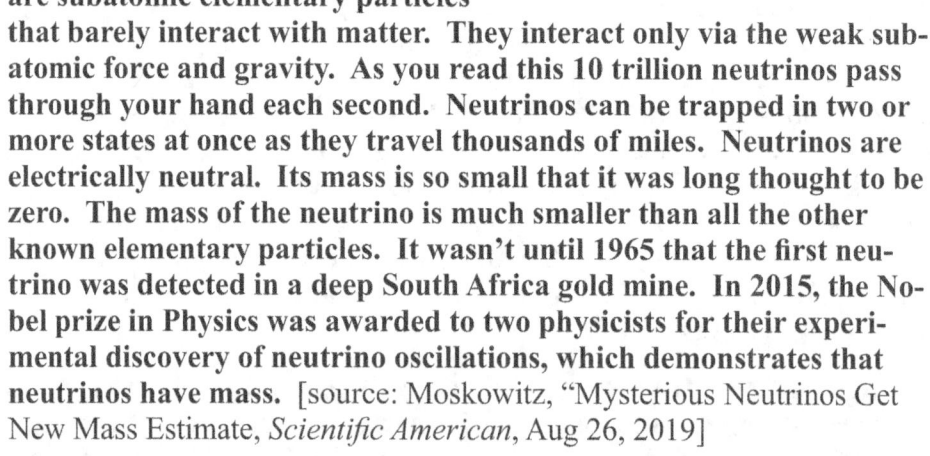

Neutrinos. The ghostly particles called neutrinos act oddly. Neutrinos are subatomic elementary particles that barely interact with matter. They interact only via the weak subatomic force and gravity. As you read this 10 trillion neutrinos pass through your hand each second. Neutrinos can be trapped in two or more states at once as they travel thousands of miles. Neutrinos are electrically neutral. Its mass is so small that it was long thought to be zero. The mass of the neutrino is much smaller than all the other known elementary particles. It wasn't until 1965 that the first neutrino was detected in a deep South Africa gold mine. In 2015, the Nobel prize in Physics was awarded to two physicists for their experimental discovery of neutrino oscillations, which demonstrates that neutrinos have mass. [source: Moskowitz, "Mysterious Neutrinos Get New Mass Estimate, *Scientific American*, Aug 26, 2019]

Photoelectric Effect. The photoelectric effect is the emission of electrons when electromagnetic radiations, like light, hits a material. You would think that the photoelectric effect is attributed to the transfer of energy from the light to an electron. In other words, the intensity of light should induce more kinetic energy. The brighter the light,

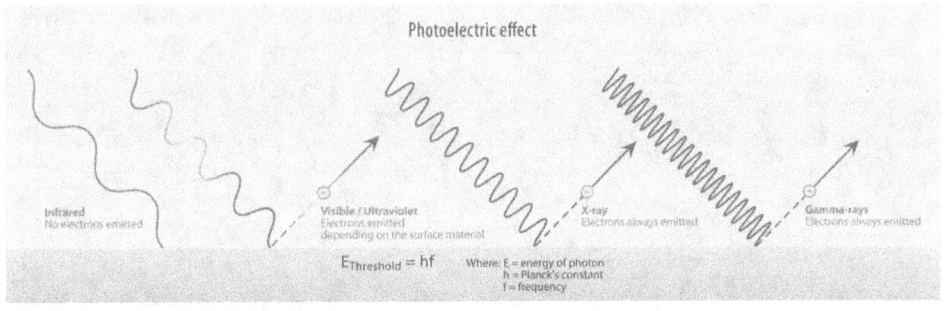

the more energy. But that is not the case. Instead, experiments showed that electrons are dislodged only by the impingement of light when it reaches a certain frequency, not intensity. Light with energy above a certain point can be used to knock electrons loose, freeing them from a sold metal surface. Ultraviolet light has a greater frequency than visible light or infrared, so it causes more energy to knock loose the electrons, not a brighter light in the visible spectrum. In 1905, Albert Einstein published a paper advancing the hypothesis that light energy is carried in discrete quantized packets to explain the photoelectric effect. In 1914, Robert Millikan designed an experiment that supported Einstein's model of the photoelectric effect. In 1922, Einstein was awarded the Nobel Prize in Physics for his discovery of the law of the photoelectric effect. Millikan was awarded the Nobel Prize in 1923 for his work on the photoelectric effect. [source: Howell, "Photoelectric Effect: Explanation & Applications," *Live Science*, Apr 25, 2017]

Quark. A quark is a type of elementary particle and a fundamental constituent of matter. They are the basis of subatomic particles called hadrons. The most stable of hadrons are the protons and the neutrons. Physicists refer to different types of quark as 6 different flavors: up, down, strange, charm, bottom, and top. The quarks are the smallest objects of what has been monitored to the present day. There is no evidence that quarks are made up of smaller particles. Quarks are one of the first particles formed for the universe. The quark model was independently proposed by Murray Gell-Mann (1929-2019) and George Zweig (1937-) in 1964. Gell-Mann won the Nobel Prize in Physics in 1969 for his contributions and discoveries concerning the classification of elementary particles. Zweig was nominated for a Nobel Prize, but never won one. [source: Howell, "7 Strange Facts About Quarks," *Live Science*, May 5, 2014]

Bill Wall

Oddities of Science

Oddities in Zoology

Bermuda petrel. The Bermuda petrel (Cahow) is a seabird. It is the national bird of Bermuda and can be found pictured on Bermudian currency. It is the second rarest seabird on the planet, behind the Balearic shearwater seabird. For over 300 years, it was thought to be extinct. In 1951, 18 nesting pairs of the Bermuda petrel were found. Special glands in their tube-like nostrils allow them to ingest seawater. These glands filter out the salt and expel it through sneezing. [source: DeVos, "Rare Birds, The Extraordinary Tale of the Bermuda Petrel and the Man Who Brought It Back from Extinction," *audubon.org*, May 17, 2013]

Camel. Camels do not store water in their humps. Camels can drink 30 gallons of water in just 13 minutes. The water is stored in their bloodstream, rather than being stored in its fatty hump, which is a source of nourishment when food is scarce. One-humped camels make up 94% of the world's camel population. The two-humped camel makes up 6%. Camels are the only hoofed animal that mates in a sitting position. Unlike other mammals, the red blood cells of a camel are oval rather than circular in shape. [source: Bradford, "Camels: Facts, Types & Pictures," *Live Science*, July 11, 2017]

Cheetah. The cheetah is capable of accelerating up to 70 miles per hour on short distances of 330 feet. This makes the cheetah the fastest land animal on Earth. For longer distances, cheetahs run at a speed of 40 miles per hour while hunting. The foot pads of a cheetah are hard and less rounded than other cats. These pads function like tire treads, providing them with increased traction in fast, sharp turns. Cheetahs are Africa's most endangered big cat. [source: "About Cheetahs," *cheetah.org*, June 25, 2019]

Coelacanth. The coelacanth (Latimeria chalumnae or "hollow spine") is a fish species thought to have become extinct 66 million years ago during the Cretaceous-Paleocene extinction event. It was initially known only from 80-million-year-old fossils. However, in De-

cember 1938, a living population was discovered in the western Indian Ocean off the east coast of South Africa at the mouth of the Chalumna River. In 1997, 6,000 miles away from the first, another population was found near the island of Sulawesi in Indonesia. The coelacanth is considered a critically endangered species. Between 1938 and 1975, only 84 specimens have been caught and recorded. The skull of this fish is split in half by a special 'intracranial joint' and its brain in only 1% the size of the cavity that houses it. [source: "Fish that outlived dinosaurs reveals secrets of ancient skull evolution," *EurekAlert*, April 17, 2019]

Cuttlefish. Cuttlefish are marine mollusks, related to snails and clams. They are odd because they have three hearts and have 8 arms and 2 tentacles. They have the largest brain-to-body size ratios of all invertebrates (animals without a backbone). That helps it to learn and remember. Their eyes allow them to see forward or backward. The blood of a cuttlefish is blue-green. It uses copper-containing protein instead of red, iron-containing protein to carry oxygen. It has tentacles attached to its head. Cuttlefish produce clouds of ink when they feel threatened. The ink use to be used by artists and writers. **Cuttlefish propel itself backwards through the water with 'jet propulsion.' Wild cuttlefish are not found in U.S. waters.** [source: Kennedy, "10 Cuttlefish Facts," *ThoughtCo.,* Jan 26, 2019]

Falcon. Falcons are the most widespread birds of prey. There are about 40 species of falcons. Their vision is 2.6 times that of a normal human. They are the fastest-moving creatures on Earth. The fastest recorded dive of a falcon was 242 miles per hour. Falcons are more closely related to parrots than to hawks or eagles. [source: Gwin, "Inside a Sheikh's Plan to Protect the World's Fastest Animal," *National Geographic*, October 2018]

Oddities of Science

Hammerhead shark. The hammerhead shark is so named for its unusual and distinctive structure of their heads, but their heads don't really resemble a hammer. The hammerhead shark is the oddest of all sharks. They are the youngest extant species of sharks that go back 20 million years. The first shark species appeared 450 million years ago. The positioning of their eyes, mounted on the sides of the shark's head, allows 360 degrees of vision in the vertical plane, meaning the hammerhead sharks can see above and below them at all times. They can grow to 20 feet long and weigh over 900 pounds. There has never been a recorded human fatality from this species of shark. [source: "Great Hammerhead Shark," Oceana, March 29, 2019]

Hummingbird. Hummingbirds are the smallest of birds. The smallest hummingbird was 2 inches long and weighed less than 2 grams (0.07 ounces), not even half the weight of a nickel. They are known as humming-birds because of the humming sound created by their beating wings, which flap at high frequencies that can be heard by humans. The hummingbird is the only bird that can fly backwards. In one day, it uses in proportion to its size, a little over 44 times as much energy as an average man. Its nest is the size of a thimble and is the smallest built by any bird. It lays the smallest egg, less than ½ inch long. There are over 350 known species of hummingbirds, all which live in the Americas. Two species have already gone extinct in the 20th century. The sword-billed hummingbird has a beak that is longer than its body. A hummingbird's heart rate has been measured at 1,260 beats per minute. Such a high metabolism means hummingbirds have to consume several times their weight in nectar per day, or they will starve to death. The oldest hummingbird in captivity lived for 17 years. [source: Coppard, "11 incredible hummingbird facts," *Discover Wildlife*, May 23, 2019]

Inland Taipan. The Inland Taipan (also called Western Taipan) of central east Australia is the most venomous snake ever found. Its venom could kill 125 people. Oddly, there has never been a human fatality from it, probably owing to its remote habitat. In 1879, it was first described by the Irish zoologist Sir Frederick McCoy (1817-1899), who was active in Australia. It was next seen in 1882, but for some odd reason, the snake was not seen for 90 years and no specimen was ever found. The snake disappeared from taxonomic literature and guides to Australia's snakes, Then, in August 1972, it was rediscovered. In 1974, the first live specimens were available to taxonomists and toxinologists. [source: Pultarova, "Pet Snake Nearly Kills Teen: Why the Inland Taipan Is So Deadly, *Live Science*, Nov 9, 2017]

Komodo dragon. The Komodo dragon is a species of lizard found in the Indonesian island of Komodo and four other nearby islands. It is the largest extant species of lizard that can grow as long as 10 feet and weigh up to 150 pounds. They can eat up to 80% of their own  body weight in one sitting. They can smell blood 6 miles away. They were first recorded by Western scientists in 1910. Islanders called them "land crocodiles." In 1927, the first live Komodo dragons were put on display at the Reptile House at the London Zoo. In 1926, an expedition led by Douglas Burden to Komodo Island brought back 12 preserved Komodo dragon specimens and 2 live ones. The scientific expedition provided the inspiration for the 1933 movie *King Kong*. It was Burdon who coined the common name "Komodo dragon." In 2007, a Komodo dragon killed an 8-year-old boy on Komodo Island, marking the first fatal attack on a human being in 33 years. In 2009, a man set out to gather apples from an orchard on Komodo Island. He slipped and fell from an apple tree where two Komodo dragons were waiting below. He suffered fatal injuries from the Komodo dragon attack. [source: Nuwer, "The Most Infamous Komodo Dragon Attacks of the Past 10 Years, *Smithsonian Magazine*, Jan 24, 2013]

Millipede. A white millipede called Illacme plenipes is the world-record holder for the "leggiest creature." It has up to 750 wiggly legs, more than any other animal in the world, on a body about 1 inch long. It is only found in a small area in San Benito County in northern California. That's odd because its closest living relative is located in South Africa. The species was first discovered in 1926, rediscovered in 2005. A new species, Illacme tobini, was discovered in 2016. [source: Andries, "World's Leggiest Animal Found Near Silicon Valley," *National Geographic*, Nov 12, 2012]

Nematode. Nematodes are roundworms with tubular digestive systems with openings at both ends. There are over 25,000 known species and it may be as many as 40,000 species worldwide. There are 60 billion nematodes for each human. They make up 80% of all individual animals on earth. The smallest nematodes are microscopic. The largest parasitic species are up to 3 feet in length. Nematodes can be found living 0.8 miles below Earth's surface, deeper than any other multicellular animal. Certain nematodes that live on an island in the Indian Ocean can develop one of 5 different mouths, depending on what type of food is available. In Siberia, melting permafrost is releasing microscopic nematodes that have been frozen or "asleep" for 42,000 years. Two species of these worms were successfully revived. After they were defrosted, researchers saw them moving and eating. This represents the first evidence of multicellular organisms returning to life after a long-term slumber. [source: "Viable Nematodes from Late Pleistocene Permafrost of the Kolyma River Lowland," *Doklady Biological Sciences*, July 16, 2018]

Okapi. The okapi (pronounce oh-COP-ee) is an odd animal that has white and black stripes on its legs like a zebra, a body of a horse, and a head that looks like a giraffe. It is the only living relative of the giraffe. The animal was not discovered until 1901. It is native to the northeast area of the Congo in Central Africa. It has a body length of over 8 feet, about 5 feet tall, weighing almost 800 pounds. There are about 25,000 okapis in the wild. [source: Bradford, "Okapi: Facts About the Forest Giraffe," *Live Science*, Sep 23, 2016]

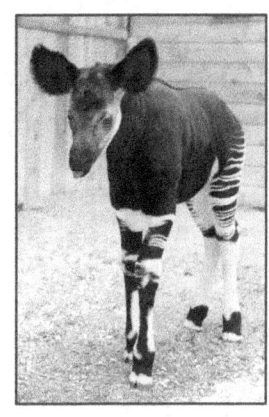

Panda. The panda is a bear only native to south central China. Bamboo shoots and leaves make up more than 99% of its diet. Male pandas can weigh up to 350 pounds, while females can weigh up to 150 pounds. A panda's paw has a thumb and 5 fingers. Initially, the primary method of breeding pandas in captivity was by artificial insemination, as they lose interest in mating once they are captured. Scientists have also tried special herbs, Viagra, and even showing videos of pandas mating, popularly known as "panda porn." A panda presses its head briefly against a tree trunk as it climbs. The head serves as a make-do extra paw. The only other animal that uses its head to climb is the baby kangaroo as it hauls itself to its mother's pouch for the first time. [sources: Milius, "Pandas use their heads as a kind of extra limb for climbing," *Science News for Students,* March 13, 2020 and Alsop, "Panda porn shown in attempt to get two to mate," *The Telegraph* (UK), March 15, 2013]

Rabbit. A rabbit begins to breed when it is 6 month old. It then can have 4 to 8 litters a year, each bearing 3 to 8 baby rabbits. A rabbit usually lives 7 to 8 years. In 1859, a few pairs of rabbits were released in the wild in Australia by an English hunter, Thomas Austin. Within

10 years, there were many millions of rabbits in Australia. It was the fastest spread ever recorded of any mammal anywhere in the world. Rabbits in Australia are the most significant known factor in species loss in Australia. Currently, there are over 200 million rabbits that inhabit 2.5 million square miles of Australia. In the 20th century, there were 600 million rabbits in Australia. [source: "How Rabbits Took Over Australia," *All That's Interesting*, June 13, 2019]

Seals. Marine animals like seals (pinnipeds) usually communicate vocally using calls, barks, and whistles. A recent study has discovered that grey seals can also clap their flippers underwater during the breeding season, as a show of strength that warns off competitors and advertises to potential mates. There are 33 extant species of seals and over 50 extinct species that have been discovered. Though not as fast in the water as dolphins, seals are more flexible and agile. In 1971, a captive harbor seal was trained to imitate human words, phrases, and laughter. [sources: "Grey seals discovered clapping underwater to communicate," *Science Daily*, Feb 3, 2020 and "Percussive underwater signaling in wild grey seals," *Marine Mammal Science*, 2020]

Oddities of Science

Squid. A squid had 8 arms and two tentacles. The giant squid is the largest invertebrate (no backbone) on the planet. It is not a fish. The two long tentacles are used to grab prey and the 8 arms are used to hold and control it. The beak then cuts the food into smaller chunks to 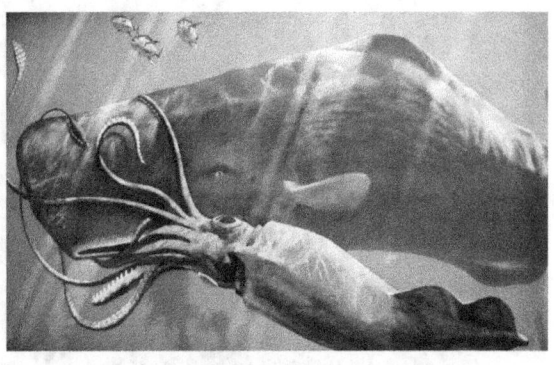 swallow. Squids are among the most intelligent of invertebrates. They hunt cooperatively. They can change color or eject a cloud of ink to distract predators. The largest giant squid ever recorded measured 47 feet in length and weighed nearly 2,000 pounds. In 2007, the largest squid was caught off the coast of Antarctica weighing over 1,000 pounds and measuring 33 feet. In 2012, scientists filmed a giant squid in its natural habitat for the first time. The giant squid lives up to 3,300 feet below the ocean's surface. The eyes of a squid can exceed the size of dinner plates, about 1 foot in diameter, making them the largest eyes in the animal kingdom. There are an estimated 500 species of squid. Giant squids live for about 5 years and reproduce only once. Its only enemy in the sea is the sperm whale. Squids never stop growing. The longer they live, the bigger they become. [source: Roper, "Giant Squid," *Smithsonian Ocean*, April 2018]

Squirrel. Squirrels are in a family that includes rodents. There are 280 different species of squirrel. Tree squirrels can descend a tree head-first because they can rotate their ankles 180 degrees, enabling the hind paws to point backward and thus grip the tree bark from the opposite direction. Squirrels can find food buried beneath a foot of snow. A squirrel's front teeth never stop growing. This ensures that their teeth don't wear down from constant gnawing on nuts and other objects. Squirrels zigzag to avoid predators. Squirrels are born blind and toothless. Squirrels cannot digest cellulose. Squirrels are not capable of vomiting or burping. [source: "10 nutty facts about squirrels," Discover Wildlife, June 19, 2018]

Whale. The blue whale is the largest animal known to have existed. They can grow to 105 feet in length and weigh up to 200 tons. All it eats is krill, the tiny plankton found floating in the water of the oceans. During certain times of the year, a blue whale can consume about 4 tons of krill a day. The blue whale can go for 6 months with-

out eating, surviving on its blubber for all that time. A whale can live up to 90 years. Within a year from the time a blue whale is newly conceived as a fertilized egg weighing a few ounces, it becomes a young blue whale weighing 49 tons. This means that it has multiplied its starting weight by 300 billion times. This is a faster rate of growth than any other living creature. Its tongue weighs more than an elephant. Its heart is the size of a car. A blue whale's heart beats about 9 times a minute. [source: "Blue Whale," National Geographic, Feb 18, 2017]

Zebra. Zebras are three species of African equids (horse family) united be their distinctive black-and-white striped coats. It was once thought that zebras were white animals with black stripes, since some zebras have white underbellies. New evidence, however, shows that the animal's background color is black, and the white stripes and bellies are additions. Scientists are not sure of the evolution of the stripes. They may be related to camouflage, but the current leading hypothesis is that the stripes confuse the vision of biting flies, such as the blood-sucking tsetse flies and horseflies. In February 2019, UK scientists performed an experiment putting striped blankets on horses, and found that most flies were unable to make a controlled landing on the striped patterns. Stripes on a zebra are unique as fingerprints – no two are  exactly alike – although each of the three species has its own general pattern. Stripes may also help zebras recognize one another and help in thermal regulation. [source: Law, "The truth behind why zebras have stripes," *BBC*, Dec 11, 2019]

Oddities of Science

Oddities of Discovery

Accidental Discoveries

Cosmic Microwave Background Radiation. In 1964, Arno Penzias (1933-) and Robert Woodrow Wilson (1936-) were trying to make a new type of supersensitive 20-foot horn antenna owned by Bell Laboratories in New Jersey. But they were getting a troubled persistent background noise. It was a steady hiss that made their experimental work impossible. It came from every direction, day or night, for a year. The scientists tried everything to eliminate the noise, including scrubbing the inside of the device from all the pigeon droppings. The scientists found a source of noise in the sky that they would not explain. They finally identified the noise as Cosmic Microwave Background (CMB) radiation that was left over from the Big Bang, or the time when the universe began. Although Penzias and Wilson were not looking for CMB, didn't know what is was when they found it, did not give it a good description or interpret their findings in any paper, they received the 1978 Novel Prize in Physics. The team that was looking for CMB, Robert Dicke, Jim Peebles, and David Wilkinson, astrophysicists at Princeton University, 37 miles away, did not get the Nobel Prize. [source: Howell, "Cosmic Microwave Background: Remnant of the Big Bang," *space.com*, Aug 24, 2018]

Microwave Oven. In 1945, Percy Spencer (1894-1970), a physicist and engineer for Raytheon Corporation was working on a radar-related project and building magnetrons. While testing a new vacuum tube and standing in front of an active radar set, he discovered that a chocolate bar he had in his pocket melted. He soon experimented by aiming the tube at other edible items, such as eggs and popcorn kernels. Heat was cooking these items and he created the world's first microwave popcorn. When he aimed the tube at an egg placed in a tea kettle, the egg exploded in the face of one of his co-workers. Spencer concluded that the heat the objects experience was

from the microwave energy. Soon after, Raytheon patented the first microwave oven, called Radarange. It wasn't until 1967 that the first relatively affordable microwave oven was available for sale. For his invention, Spencer received no royalties. Raytheon paid him a one-time gratuity check for $2.00 for his invention. [source: Ross, "Who Invented the Microwave Oven?" *Live Science*, Jan 6, 2017]

Play-Doh. Play-Doh is a non-toxic, non-staining, modeling compound for kids. It was accidently discovered by Joseph McVicker (1929-2011) and his uncle, Noah McVicker (1905-1980), who were working on creating a wallpaper cleaner for their soap company, Kutol. The company's primary product – a soft, pliable compound used for wiping soot from wallpaper – was no longer in demand. The McVickers' soon realized that their wallpaper cleaner could be used as a pliable modeling clay. The product was tested in Cincinnati-area schools and daycares. The McVickers' applied for a patent in 1958, but the U.S. Patent Office did not officially patent the clay until 1965. [source: Kindy, "The Accidental Invention of Play-Doh," *Smithsonian*, Nov 12, 2019]

Slinky. A Slinky is a precompressed helical spring toy. In 1943, Robert James (1914-1974) was a mechanical and naval engineer at the Philadelphia naval dockyards who was trying to develop a means for suspending sensitive shipboard instruments aboard naval vessels, even in rough seas. As he was working on some tension springs, he accidently dropped one. Seeing how the spring kept moving after it hit the ground, the idea for a toy was born. In December 1945, James took his entire inventory of 400 Slinky toys to the Gimbals department store in Philadelphia and sold his entire inventory at $1 a pop ($14 in today's currency) in 90 minutes. James and his wife Betty got a $500 loan, and then formed James Industries to manufacture the product. They patented their product in 1946. [source: Lallensack, "The Accidental Invention of the Slinky," *Smithsonian Magazine*, Aug 19, 2019]

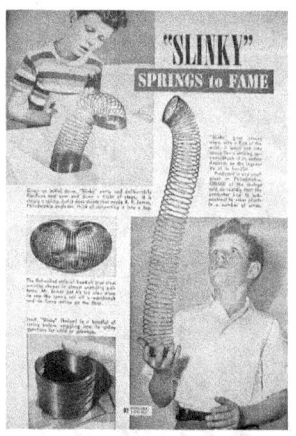

Oddities of Science

Super Glue. In 1942, Dr. Harry Coover (1917-2011), an organic chemist at the B. F. Goodrich Company, created a substance called cyanoacrylate. It was supposed to be used to make clear plastic gun sights, but it was too sticky to be of use, so he set it aside. Nine years later, in 1951, while working at the Eastman-Kodak Laboratory, he was investigating heat-resistant polymers for jet canopies  and was again experimenting with cyanoacrylate. Once again, it proved too sticky to be used as a jet canopy. Then one day, a chemist, Fred Joyner, in his group informed Coover that he had permanently damaged an expensive refractometer by gluing it together. This time, Coover observed the stuff formed an incredibly strong bond without needing heat. Coover discovered a unique adhesive. In 1958, Kodak marketed the product as Eastman 910. During the 1960s, Eastman Kodak sold cyanoacrylate to Loctile, which repackaged and distributed it as Locite Quick Set 404, then Super Bonder, then LociteSuper Glue. Coover held over 460 patents when he died in 2012, but thought that Super Glue was his best discovery. [source: Grossman, "The Serendipitous History of Superglue," *Popular Mechanics*, Feb 6, 2017]

Sweet'N Low. In 1878, chemist Dr. Constantine Fahlberg (1850-1910) was working an analysis of coal tar in a lab. When he returned home, he forgot to wash his hands before dinner, and started eating his food. He noticed that everything tasted sweet on everything he touched. He returned to the lab and started testing compounds, tasting random chemicals and the contents of every beaker and evaporating dish on the lab table (not recommended). He finally found that combining sulfobenzoic acid with phosphorus chloride and ammonia created a substance that was 400 times sweeter than sugar. He had created anhydroorthosulphaminebenzoic acid. Fahlberg then patented his discovery and called it saccharin, which is the artificial

 sweetener in Sweet'N Low, a zero calorie sweetener. In 1912, the sweetener was banned because it was thought to be toxic. The ban was later reversed and the artificial sweetener became widespread during World War I when sugar was rationed. [source: Gershenson, "The bizarre history of artificial sweeteners," *Fox News*, March 8, 2017]

Teflon. Teflon, also known as polytetrafluoroethylene (PTFE), was accidently discovered in August 1938 by Dr. Roy Plunkett (1910-1994) while he was a research chemist at DuPont. As he was attempting to make a new refrigerant, the gas in its pressure bottle stopped flowing before the bottle's weight had dropped to the point signaling "empty."

He became curious as to the source of the weight, so he sawed the pressure bottle in half. He found the bottle's interior coated with a waxy white material that was oddly slippery. The substance had amazing properties such as resistance to corrosion, low surface friction, and high heat resistance. The product was patented in 1941 and was registered with the Teflon trademark in 1945. An early use of Teflon was in the Manhattan Project as a material to coat valves and seals in the pipes holding highly reactive uranium hexafluoride. In 1954, it was used to create non-stick pans in France. It was introduced to the U.S. market in 1961. Today, 70% of all cookware sold in the United States is nonstick. [source: Bellis, "The Invention of Teflon: Roy Plunkett," *ThoughtCo*, Jan 20, 2020]

Uranus. Uranus, the 7th planet from the Sun, was discovered completely by accident. It had been observed on many occasions before its recognition as a planet, but it was generally mistaken for a star. In March 1781, British astronomer Sir William Herschel (1738-1822) observed it from the garden of his home in Bath, England, but he thought it was a comet. The Finnish-Swedish astronomer Anders Lexell (1740-1784), working in Russia, was the first to compute the orbit of the new object. Its nearly circular orbit led him to a conclusion that this was a planet, not a comet. Herschel and others now realized it was a new planet in orbit around our sun, the first new planet discovered since ancient times. It took 70 years before the name of the planet, Uranus, was accepted. Herschel named it Georgium Sidus (George's Star) or the "Georgian Planet" in honor of King George III. Herschel's proposed name was not popular outside Britain, and alternatives were soon proposed. The astronomer Johann Bode (1747-1826) proposed Uranus, the Latinized version of the Greek god of the sky, Ouranos. It is the only planet to be named after a Greek god rather than a Roman one. That name finally became universal in

Oddities of Science

1850 when it was accepted in Her Majesty's Nautical Almanac. [source: Choi, "Uranus: The Ringed Planet That Sits on its Side," *space.com*, July 10, 2019]

Vaseline. Vaseline was not discovered by a scientist, medical doctor, or engineer. It was discovered by oil rig workers. In 1859, petroleum was discovered near Titusville, Pennsylvania. Soon, oil fields were becoming very prolific. But the oil rig workers quickly discovered a problem. An oily wax tended to accumulate on their rigs, and the residue caused the machinery to malfunction. During the 1850s, an American chemist, Robert Chesebrough, was clarifying kerosene from sperm whales. The discovery of petroleum in Titusville rendered his job obsolete. So, he traveled to Titusville to research what new materials might be created from the new fuel. Upon visiting Titusville, Chesebrough heard about the "rod wax" the oil workers 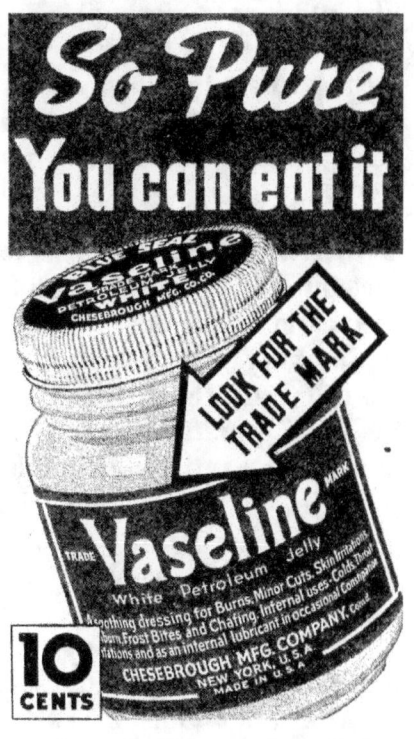 had to deal with. He also heard that the workers had started applying the wax to their wounds, finding that it helped them heal faster. Chesebrough decided to take some wax back to his lab in New York. There, he refined it into a gel, patented the process for making it, and started selling it under the new name Vaseline. Chesebrough then traveled around New York to demonstrate his new wonder product. He demonstrated the power of Vaseline by burning himself with acid or an open flame and applying the Vaseline to show its healing effects. Chesebrough lived to 96, crediting Vaseline for his long life. He ate a spoonful of Vaseline every day. [source: Jayakumar & Micheletti, "Robert Chesebrough and the Dermatologic Wonder of Petroleum Jelly," *JAMA Network*, November 2017]

Velcro. In 1941, Swiss electrical engineer George de Mestral (1907-1990) went for a hike in the Alps. When he returned home, he noticed all these burdock burrs (seeds) stuck to his clothes. Under a microscope, he discovered that the little seeds were covered in hundreds of small hooks, which caught on anything with a loop, such as clothing, fur, or hair. After noting how firmly these burrs attached to fabric, he decided to invent the hook and loop fastener, which he named Velcro, from the French words velours (velvet) and crochet (hook). It took him 10 years to mechanize the process that worked. He patented Velcro in 1955. The first notable use of Velcro came in the aerospace industry, where it helped astronauts get in and out of bulky space suits. Velcro was used during the Apollo missions to anchor equipment in zero gravity. [source: Daniels, "The Invention of Velcro," *ThoughtCo*, Jan 23, 2020]

Viagra. In 1995, chemists work at Pfizer in Sandwich, Kent, introduced the chemical sildenafil, the active drug in Viagra, as a heart medication. The drug was initially studied for use angina pectoris (a symptom of heart disease) and high blood pressure. However, during clinical trials, the drug proved ineffective for heart conditions. However, the men in the clinical trials all noted that the medication had another effect – stronger and longer-lasting erections. Pfizer conducted clinical trials on 4,000 men, and saw similar results. Pfizer therefore decided to market it for erectile dysfunction. The drug was patented in 1996 and approved by the FDA in 1998. It was the first-ever oral treatment of erectile dysfunction (ED). Annual sales of Viagra are about $2 billion a year. Over 62 million men over the world have bought the drug. Levitra and Vivanza are brand names for vardenafil, and Cialis is a brand name for tadalafil. [source: Foley, "How Viagra was discovered by Pfizer," *Quartz*, Sep 10, 2017]

Vulcanized rubber. In 1839, Charles Goodyear (1800-1860), while at the Eagle India Rubber Company in Massachusetts, was looking for a way to fix the current flaws of rubber, which solidified and cracked in winter, and melted in the summer heat. Goodyear discovered vulcanized rubber by accident when he happened to spill a mixture of rubber, sulfur, and lead on a hot stove, which caused the rubber to vulcanize. He patented his vulcanization process in 1844. He spent the rest of his life fighting patent infringements in courts in the US and Britain. He died at age 59 $200,000 in debt. In 1898, the Goodyear Tire & Rubber Company was founded in Akron, Ohio, and named after Goodyear by Frank Seiberling. [source: Somma, "Charles Goodyear and the Vulcanization of Rubber," *Connecticut History*, Dec 29, 2014]

Oddities of Science

X-Rays. In November 1895, German physicist Wilhelm Conrad Roentgen (1845-1923) was working with a cathode ray tube. Despite the tube being covered with heavy black cardboard, Roentgen noticed a fluorescent screen three feet away was glowing. 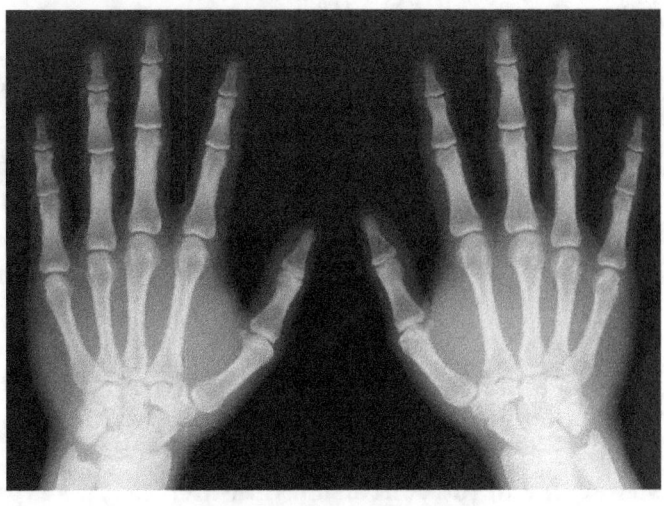 The rays were somehow illuminating the screen. When he placed his hand in front of the tube, he noticed that he could see his bones in the image that was projected on a screen. He replaced the screen with a photographic plate to capture the images, creating the first X-ray photograph. He named it X-radiation to signify an unknown type of radiation. Roentgen then took a picture of his wife's hand on a photographic plate formed due to X-rays. It was the first photograph of a human body part using X-rays, which showed her bones and her wedding ring. When she saw the picture, she said, I have seen death." His accidental discovery revolutionized the fields of physics and medicine. [source: Waters, "The First X-ray, 1895," *The Scientist*, June 30, 2017]

Odd Discoveries

Birth control pills. In 1994, scientists found that women who take birth control pills blink 32% more often than women who do not take birth control pills. In 2013, a scientific study found that women who used birth control pills for 3 years or more could be twice as likely to develop glaucoma with women who did not use birth control pills. The FDA approved the birth control pill in 1960. However, it wasn't until 1965 that contraceptives were available for married women in all states. It wasn't until 1972 that unmarried women in all states had access to the birth control pill. *Time* magazine put the birth control pill on its cover in April 1967. [source: "The effects of gender and birth control pill use on spontaneous blink rates," *Journal of the American Optometric Association*, volume 11, issue 11, 1994 pp. 63-770]

Gravitational waves. Gravitational waves are disturbances in the curvature of spacetime, generated by accelerated masses that propagate as waves outward from their source at the speed of light (186,000 miles per second). In 1893, Oliver Heaviside (1850-1925) discussed the possibility of gravitational waves as an analogy in gravitation and electricity. In 1905, Henri Poincare (1854-1912) proposed gravitational waves. Einstein came to the conclusion that there were 3 types of gravitational waves, but he later doubted the result from his calculations. Einstein's Theory of General Relativity predicted that mass actually curves space-time, but it wasn't until 2016 that scientists were able to confirm it. In 2015, the first direct detection of gravitational waves were made by LIGO (Laser Interferometer Gravitational-Wave Observatory). LIGO measured a signal that originated from the merger of two black holes. In 2017, Rainer Weiss, Kip Torne, and Barry Barish were awarded the Nobel Prize in Physics for their role in the detection of gravitational waves. [source: "What is a Gravitational Wave?" *spaceplace.nasa.gov*, Jan 31, 2017]

Grid Cells. In 2005, scientists discovered grid cells in the entorhinal cortex of the brain. They are neurons that allows us to map out space. The neurons fire at regular intervals as an animal navigates an open area, allowing it to understand its position in space by storing and integrating about location, distance, and direction. Notably, the grid is self-generated. The cells fire almost as if they were invisible. The discoverers, Edvard Moser and wife, May-Brit Moser, were awarded the 2014 Nobel Prize in Physiology or Medicine, alongside John O'Keefe. What's odd is that grid cell activity does not require visual input, since the grid patterns remain unchanged while all the lights in an environment are turned off. [sources: "Microstructure of a spatial map in the entorhinal cortex," *Nature*, June 19, 2005 and Moser, "Place cells, grid cells, and the brain's spatial representation system," *Annual Review Neuroscience*, 2008]

Oddities of Science

Higgs boson. The Higgs boson is an elementary particle produced by the quantum excitation of the Higgs field, one of the fields in particle physics theory. Everything that has mass gets it by interacting with the Higgs field, which occupies the entire universe. It is named after physicist Peter Higgs (1929-) who proposed that broken symmetry in electroweak theory could explain the origin of mass of elementary particles. The particle is sometimes called the "God Particle." Detecting the particle was difficult to do because of the energy it took to collide particles. In July 2012, the particle was discovered in the Large Hadron Collider at CERN in the mass region of 125.35 billion electron volts (125.35 Gev). Higgs was awarded the 2013 Nobel Prize in Physics for his theoretical predictions, along with Francois Englert. [source: Irving, "CERN precisely measures the mass of the Higgs boson," *New Atlas*, Oct 27, 2019]

Quasicrystal. A quasiperiodic crystal, or quasicrystal, is a structure that is ordered but not periodic. It does not repeat its pattern. Normal crystals are restricted to 2, 3, 4. Or 6-fold rotational symmetries and are arranged in a repeating pattern. Quasicrystals shows sharp peaks with other symmetry orders, such as 5-fold and 10-fold rotational symmetries. In 2009, natural quasicrystals were discovered. The 2011 Nobel Prize in Chemistry was awarded to Israeli materials scientist Dr. Dan Shechtman (1941-) who discovered quasicrystals. Shechtman published his findings of quasicrystals in 1984, but the interpretations for them were not accepted by many in the scientific community. Nobel Laureate Linus Pauling called him a fraud, saying that "There is no such thing as quasicrystals, only quasi-scientists." [source: Glotzer, "Quasicrystals: the thrill of the chase," *Nature*, Jan 8, 2019]

Sliced Bread. In 1927, American inventor and optics engineer Otto Rohwedder (1880-1960) of Davenport, Iowa, created the first automatic bread-slicing machine for commercial use. The machine not

only sliced the bread but wrapped it. His skeptics thought factory-sliced loaves would quickly go stale or fall apart. Rohwedder patented his "power-driven, multi-bladed" machine and sold the first machine to a friend and baker in Chillicother, Missouri, in July 1928. After 6 months of use, the machine fell apart. In 1930, the Continental Baking Company introduced pre-sliced Wonder Bread using machines that were improved versions of Rohwedder's design. By 1933, American bakeries for the first time produced more sliced bread than unsliced bread loaves. During World War II, sliced bread was briefly banned by the U.S. government in an effort to conserve resources, such as the paper used to wrap each loaf to help maintain freshness. Also, there was a wartime conservation measure meant to save the 100 tons of steel that went into slicing machines each year. The ban on sliced bread provoked as much ire as gas rationing did. The unpopular ban was lifted two months after it went into effect. [source: Latson, "How Sliced Bread Became the 'Greatest Thing'," *Time*, July 7, 2025]

Thermometer. The thermometer was not a single invention, but a development. The device that measured heat and cold was first called the thermoscope. Though Galileo is often said to have been the inventor of the thermometer, there is no surviving document that he produced any such instrument. In 1714, Dutch scientist Daniel Gabriel Fahrenheit (1686-1736) invented the mercury-in-glass thermometer. However, for some odd reason, he calibrated his instrument that put freezing at 30 degrees, and the temperature when the instrument was put in the mouth at 90 degrees. Work by others showed that water boils about 180 degrees above freezing point. The Fahrenheit scale was later redefined to make the freezing-to-boiling interval exactly 180 degrees. So, freezing was set at 32 degrees and boiling was set at 212 degrees. In 1742, Anders Celsius (1701-1744) proposed an odd scale with zero at the boiling point and 100 degrees at the freezing point of water. The scale that now bears his name has the degrees the other way around. Temperature for both Celsius and Fahrenheit is the same at -40 degrees. Ear thermometers are made that uses an infrared thermometer. Rectal and oral thermometers have been superseded by NTC (negative temperature coefficient) thermistors with a digital readout. Most thermometers are filled with mercury. But mercury is unsuitable for really cold conditions, as it freezes at -39 degrees Celsius. For low temperatures, thermometers use alcohol, which freezes at -114 degrees Celsius. [source: Bellis, "The History of the Thermometer," *ThoughtCo.*, May 7, 2019]

Oddities of Science
Oddities of Scientific Studies

Cadaver arms. An odd scientific study experimented with eight severed human male forearms used to punch and slap objects. Scientists were comparing the hand of an ape with that of a human and was testing the hypothesis that the hand was usable as a fist during fighting. Each severed arm was mounted on a wooden board and suspended from a pendulum so that the hand could strike a padded surface with a force balled up as a fist or as a flat palm. The researchers found that the clenched fist packed twice the force of an open-handed slap. [source: "In vitro strain in human metacarpal bones during striking," Journal of Experimental Biology, 2015]

Cannibalism. A scientific study called "Assessing the Calorific Significance of Episodes of Human Cannibalism in the Paleolithic," showed that the calorific intake of human-cannibalism diet was much lower than a traditional meat-based diet. The paper was published in *Nature* magazine on April 6, 2017 by James Cole. An earlier study claimed that a 50 kg (110 pounds) male would yield 30 kg (66 pounds) of edible muscle mass, which in turn would yield around 4.5 kg (10 pounds) of protein, or 18,000 calories. This could serve one day's protein requirement for 60 people. [source: Cole, "Assessing the Calorific Significance of Episodes of Human Cannibalism in the Paleolithic," *Nature*, April 6, 2017]

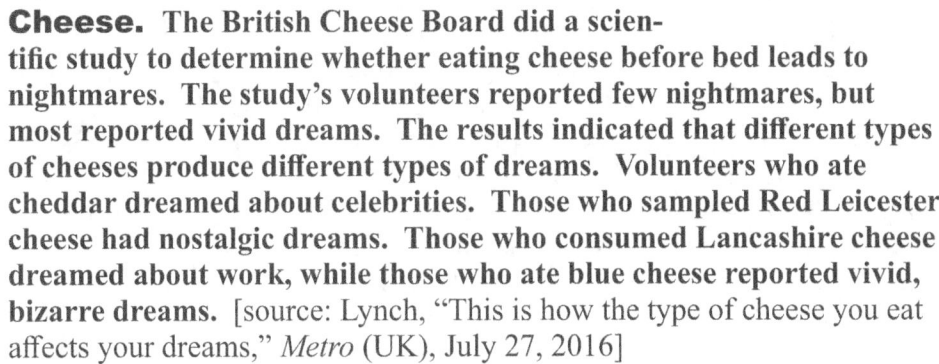

Cheese. The British Cheese Board did a scientific study to determine whether eating cheese before bed leads to nightmares. The study's volunteers reported few nightmares, but most reported vivid dreams. The results indicated that different types of cheeses produce different types of dreams. Volunteers who ate cheddar dreamed about celebrities. Those who sampled Red Leicester cheese had nostalgic dreams. Those who consumed Lancashire cheese dreamed about work, while those who ate blue cheese reported vivid, bizarre dreams. [source: Lynch, "This is how the type of cheese you eat affects your dreams," *Metro* (UK), July 27, 2016]

Clams. A scientific study was made on the affects of clams when given the antidepressant medicine Prozac. For one thing, it sends them into a breeding frenzy. [source: "Induction and potentiation of parturition in fingernail clams by selective serotonin re-uptake inhibitors," Journal of Experimental Zoology, Dec 7, 1998]

Constipated Serviceman. A scientific study called "The Constipated Serviceman: prevalence among Deployed U.S. Troops," appeared in the August 1993 issue of *Military Medicine*. The study examined the constipated rates of 500 military personnel aboard the USS Iwo Jima during its deployment during Operation Desert Shield. They found that 7% were constipated at home, 10.4% when they were aboard ship, and over 34% while in the field on military maneuvers. The authors' conclusion was that when military personnel are deployed in the field, preventive measures for constipation should be taken in account.

Cows. A scientific study was done on how humans affects the productivity of cows. It appears that a human's attitude towards animals influences their behaviors around animals. On UK farms where cows were called by name, milk yield was 258 liters higher than on farms where this was not the case. A survey of 512 dairy farmers found that 21% of farmers believed that dairy cattle were fearful of humans. 48% of respondents attributed a cow's docility to previous human contact. 90% of respondents thought that cows had feelings. 78% thought that cows were intelligent. The study suggests that UK dairy farmers have a good quality of human-animal relationship with their animals. [source: Bertenshaw & Rowlinson, "Exploring Stock Managers' Perceptions of the Human-Animal Relationship on Dairy Farms and an Association with Milk Production," *Anthrozoos*, April 28, 2015]

Dental floss. In 1990, a scientific study was done on dental floss. The findings were published in a paper called "Patient Preference for Waxed or Unwaxed Dental Floss," and published in the February 1990 issue of the *Journal of Periodontology*. 100 patients volunteered to sample a brand of similar-appearing waxed and unwaxed dental floss. Waxed floss was found to be preferred by 79% of the patients, while 21% favored 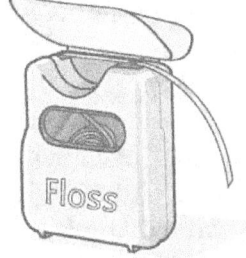 the unwaxed dental floss. The conclusion of the study was that dental professionals should recommend waxed dental floss to their patients.

Elephant and LSD. In 1962, a group of Oklahoma City researchers wanted to know what would happen if you gave an elephant

Oddities of Science

LSD. Warren Thomas, the director of the Oklahoma City Zoo, fired a cartridge-syringe containing 297 milligrams of LSD into Tusko the Elephant, "the prize of Oklahoma City Zoo." The amount was the largest dose of LSD ever given to a living creature. Tusko reacted to the shot as if a bee had stung him. He stomped around his pen a few minutes before keeling over to his side and shuddering violently. The researchers tried to revive him with Thorazine, an anti-psychotic, but after an hour and 40 minutes, the elephant was dead. The conclusion was that "It appears that the elephant is highly sensitive to the effects of LSD. [source: Pilkington, "Tusko's last trip," *The Guardian*, Feb 25, 2004]

Flea jumps. A scientific study was done comparing the flea jumps on dogs to the flea jumps on cats. The cat fleas jumped from 2 centimeters to a maximum of 48 centimeters. The dog fleas jumped from 3 centimeters to a maximum of 50 centimeters. [source: Cadiergues, "A comparison of jump performances of the dog flea, Ctenocephalides canis, and the cat flea, Ctenocephalides felis," Veterinary Parasitology, Oct 2000]

Frog odors. A scientific study was made on frog odors, with a discussion on the human perception and classification of 131 species of frogs. Volunteers were asked to smell stressed frogs and describe the odor they perceived. Some smell like mink, some smell like vanilla, and some smell like onions. The paper hopes to stimulate research into the role and chemistry identity of volatile secretions. [source: "A survey of frog odorous secretions, their possible functions and phylogenetic significance," *Applied Herpetology*, 2004]

Hamsters. A scientific study was made of hamsters, and scientists discovered that hamsters recover from jetlag more quickly when given small doses of Viagra (sildenafil). Hamsters adjusted more quickly to laboratory simulations of a 6-hour time-zone change. So, Viagra might also help people overcome jet lag faster. [source: "Sildenafil accelerates reentrainment of circadian rhythms after advancing light schedules," *Proceedings of the National Academy of Sciences*, May 22, 2007]

Hanging. In the early 1900s, Romanian forensic scientist Nicolae Minovici (1868-1941) studied the connections between tattooing and criminal behavior. He also did research on hanging and its physiological effects on the human body. He hung himself for up to 25 seconds with an assistant nearby. He experimented with different hangman's knots and the various positions of the rope around the neck. He was able to observe first-hand the symptoms of hanging, like vision problems, change in skin coloring, and ringing of the ears. He experienced pain, while swallowing, for a month following a hanging experiment. He died at age 72 from an illness affecting his vocal cords.

Knuckle cracking. In this 50-year experiment, Dr. Donald Unger, MD, tested the hypothesis that knuckle cracking leads to arthritis. During the author's childhood, various renowned authorities (his mother and several aunts, and later, his mother-in-law) informed him that cracking his knuckles would lead to arthritis of the fingers. For 50 years, he cracked his knuckles on his left hand at least twice a day, leaving those on the right hand as a control with no knuckle-cracking. Thus, the knuckles on the left hand were cracked at least 36,500 times in 50 years, while those on the right cracked rarely. At the end of 50 years, the hands were compared for the presence of arthritis. There was no evidence of arthritis in either hand, and no apparent differences between the two hands. In conclusion, there was no apparent relationship between knuckle cracking and the subsequent development of arthritis of the fingers. It should be noted that the study was entirely funded at Dr. Unger's expense, with no grants from any governmental or pharmaceutical source. [source: Unger, "Does knuckle cracking lead to arthritis of the fingers?" *Arthritis & Rheumatism*, May 28, 2004]

Leeches. A scientific study was made on to see what leeches preferred to eat or drink. From ancient times to the 19th century, leeches were used in medicine to draw blood from patients. A common practice was to immerse them in strong beer before applying them to the

Oddities of Science

patient. The new scientific study examined the use of beer (Guinness stout or Hansa bock), soured cream, and garlic on the appetite of leeches. The conclusion was that sour cream did not make the leeches hungry for blood, garlic killed them, and the beer made them drunk. [source: Baerheim & Sandvik, "Effect of ale, garlic, and soured cream on the appetite of leeches," *British Medical Journal*, Dec 24, 1994]

Mites. In 1993, Dr. Robert Lopez (1922-2007), a veterinarian from Westport, New York, infected himself by putting mites, mixed with earwax, in his ear to see what happened. He wrote up the results for the *Journal of the American Veterinary Association*. He described the sensation like insane whooshing screeching sounds. He reported it as painful, with loads of sound which got louder as the mites went deeper. By the 3rd week he had gone deaf. Then he repeated the experiment 2 more times. [source: Lopez, "Of mites and man," *Journal of the American Veterinary Medical Association*, Sep 1, 1993]

Nasal Airflow. A scientific study was made of nasal airflow and brain activity. It seems that the nose exhibits asymmetrical airflow, with the dominant airflow alternating from one nasal passage to the other over a period of hours. Based on ancient yoga breathing techniques, new evidence suggests that altering nasal airflow can influence brain activity, with reports of improved cognitive function caused by unilateral forced nostril breathing. [source: Eccles & Price, "Nasal Airflow and Brain Activity: Is There a Link?"*Journal of Laryngology and Otology*, Sep 2016]

Ostrich. A scientific study was made on the courtship behavior of adult male and female ostriches during the presence and absence of human beings. The study found that courtship behaviors in both males and females were more prevalent in the presence of humans. However, exposure to a human for a short period of time did not stimulate courtship behavior in the period immediately after the human had withdrawn. The conclusion of the study was that courtship behavior towards humans may be important in the reproductive success of ostriches in a farming environment. [source: "Courtship behavior of os-

triches towards humans under farming conditions in Britain," *British Poultry Science*, Vol 39, Issue 4, 1998]

Pigeons. A scientific study was done to show that pigeons could successfully learn to discriminated color slides of paintings by Monet and Picasso. They also showed generalization from Monet's to Cezanne's and Renoir's paintings or from Picasso's to Braque's and Matisse's paintings. Upside-down images of Monet's paintings disrupted the discrimination, whereas inverted images of Picasso's did not. The conclusion of the study suggested that pigeons' behavior could be controlled by complex visual stimuli in ways that suggest categorization. A later study showed that honeybees could discriminate between Monet and Picasso paintings. [source: "Pigeons' discrimination of paintings by Monet and Picasso," *Journal of the Experimental Analysis of Behavior*, March 1995]

Sword swallowing. A scientific study was done on sward swallowing. 110 sword swallowers from 16 countries were contacted to evaluate information on the practice and associated ill effects of sword swallowing. Sore throats were common (pretty obvious) and gastrointestinal bleeding sometimes occurred. Occasional chest pains were recorded, but most treaded without medical advice. The study did not include injuries due to swallowing glass, neon tubes, spear guns, or jack hammers (all have which had been done). One person was able to swallow 16 swords together at one time. Six sword swallowers reported perforation of the pharynx or

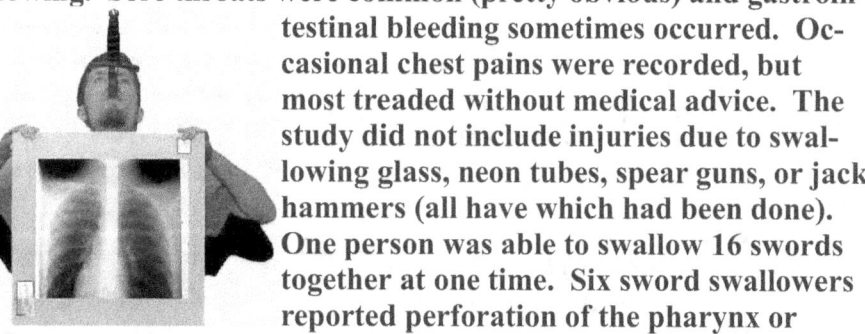

esophagus. The conclusion of the study was that sword swallowers run a higher risk of injury when they are distracted or adding embellishments to their performance. [source: Meyer & Witcombe, "Sword swallowing and its side effects," *British Medical Journal*, Dec 23, 2006]

Toast. A study was made on the dynamics of buttered toast tumbling from a table to the floor. Popular opinion is that the final state is usually butter-side down. The orthodox view is that it is random, with a 50-50 chance of landing on either side. The study showed that the toast does indeed have an inherent tendency to land butter-side down in a wide range of conditions. [source: Matthews, "Tumbling toast, Murphy's Law and the fundamental constants," *European Journal of Physics*, vol. 16, issue 4, 1995]

Oddities of Science

Woodpecker. A scientific study was made of woodpeckers, exploring and explaining why woodpeckers don't get headaches. Woodpeckers hammer their lives away for feeding, next construction, and drumming. They can strike a wooden surface of a tree of up to 20 times a second, and up to 12,000 times a day. Their pecking is equivalent to striking a wall at 16 mph – face first – each time. Woodpeckers have a thick bony skull with spongy bone and cartilage at the base of the mandible to partially cushion their blows. [source: Schwab, "Cure for a headache," *British Journal of Ophthalmology*, August 2002]

Zipper-entrapped penis. An actual scientific study called "Acute managements of the zipper-entrapped penis," appeared in the May-June 1990 issue of *The Journal of Emergency Medicine*.

This painful predicament could be made worse by overzealous intervention. The paper described a simple, basic approach to release of the penis from the zipper that was the least traumatic to both patient and provider. Since this article, other scientific studies and papers have been written, including "Unbloody Management of Penile Zipper Injury," "Penile Foreskin Trapped in a Zipper: What to Do?" and "Safe and Painless Manipulation of Penile Zipper Entrapment."

Bill Wall

Oddities of Scientific Terms

Aa. Aa is a kind of volcanic basaltic lava that forms jagged masses. Aa is the first entry in the Oxford English Dictionary. It originates from Hawaiian meaning "stony rough lava" or to "burn" or "blaze."

Arsole. Arsole is an arsenic-based organic compound. Its molecules are ring-shaped. Pure arsole, C4H4AsH, has never been isolated. [source: "Molecule of the Week Archive: Arsole," *American Chemical Society*, April 18, 2011]

Barleycorn. The barleycorn is a small English and Scottish unit of length, equal to ½ of an inch. It has been used as a measure of length since the 12th century. It was originally the length of an actual barleycorn. It is used as the basis of shoe sizes.

Barn. One barn in 10^{-28} square meters, about the cross-sectional area of a uranium nucleus. The name derives from early neutron-deflection experiments, when the uranium nucleus was described, and the phrases "big as a barn" and "hit a barn door" were used. The microbarn is called the "outhouse" and the yoctobarn is called the "shed."

Blob. A blob is a Binary Large Object. It indicates some large amount of data other than just simple text, usually stored within a database.

Bohr. The Bohr is the atomic unit of length. It is 5.291772 x 10^{-11} meters, the radius of a hydrogen atom.

Coccyx. The coccyx is the bony structure at the bottom of your spine. It is commonly referred to as the tailbone.

Cord. The cord is a unit of measure of dry volume to measure firewood and pulpwood. It is the amount of wood that can be arranged so pieces are aligned, parallel, touching and compact that occupies 128 cubic feet. This would be an area 4 feet deep by 4 feet high by 8 feet wide. Another measurement for firewood is the "rick" or face cord. It is 1/3 of a cord.

Cubit. The cubit is the distance from fingers to the elbow, or about 18 inches.

Dirac. Physicist Paul Dirac (1902-1984) was famous for saying very little as part of his personality. For information flow, one dirac is one word per hour.

Dol. The dol (from the Latin word for pain, dolor) is a unit measurement of pain. One dol is equal to "just noticeable differences" in pain. Other pain indexes are the Schmidt sting pain index and the Starr sting pain index.

FLOPS. In computing, FLOPS (Floating point operations per second) is a measure of a computer's computing power.

Formication. A medical term for the sensation that small insects are crawling over your skin. The word is derived from formica, Latin for ant.

Fukalite. A mineral composed of calcium, oxygen, and silicon. Its formula is Ca4Si2O6 (CO3)(OH,F)2.   [source: "Fukalite Mineral Data," *webmineral.com*]

Galactic year. The galactic year (GY) is the time it takes for our solar system to revolve once around the galactic core, approximately 250 million years. The age of the Earth is estimated at about 20 galactic years.

Garn. Jake Garn (1932-) was a U.S. Senator from Utah who asked to fly on the Space Shuttle because he was head of the Senate appropriations subcommittee that dealt with NASA. He flew on STS-51-D in 1985 as a payload specialist. He was the first sitting member of Congress to fly into space. He suffered from severe space sickness and vomited throughout the flight. NASA uses the Garn scale to measure space sickness. 1 Garn means a person is as sick as Jake Garn himself was. [source: Fitzpatrick, "Space Sickness is Measured In What Units?" *howtogeek.com*, Jan 8, 2019]

Googol. The number 10 raised to the power 100 (10^{100}). It is written out as the numeral 1, followed by 100 zeros. The term was coined in 1920 by 9-year-oled Milton Sirotta (1911-1981).

Hand unit. One hand unit of length is exactly 4 inches. It is used to measure the height of horses.

Oddities of Science

Hinny. A hinny is a hybrid between a male horse (a stallion) and a female donkey (a jenny).

Jiffy. A jiffy is a unit of time used in computer operating systems. It is the time between two ticks of the system timer interrupt. It is typically between 1 and 10 milliseconds. In electronics, a jiffy is the period of an alternating current power cycle, about 1/60 of a second.

KLOC. KLOC stands for kilo-lines of code, or 1,000 lines of code. I KLOC could take one programmer as long as 50 working days to complete.

Langley. The Langley (Ly) is used to measure solar radiation. It is equal to one thermochemical calorie per square centimeter.

Light-nanosecond. The light-nanosecond is defined as exactly 29.9792458 centimeters (11.80285267717 inches). It is the distance that a photon could travel in one billionth of a second. It was popularized by Grace Hopper (1906-1992), who handed out pieces of wire with that length as a visual aid of how fast the maximum speed of signals would travel in a vacuum. In many of her talks, she handed out "nanoseconds" to everyone in the audience, contrasting them to a light-microsecond, which was a coil of wire 984 feet (300 meters) long.

Mickey. One mickey is the smallest resolvable unit of distance by a given computer mouse pointing device. It is named after Walt Disney's Mickey Mouse cartoon character. A typical resolution is 500 mickeys per inch. One mickey is about 1/200th of an inch.

Micromort. A micromort is a unit of risk measuring a one-in-a-million probability of death. Smoking 1.4 cigarettes increases one's death risk by one micromort, as does traveling 230 miles by car.

Mole. A standard scientific unit for measuring large quantities of tine molecules or particles. It is the number of particles in 12 grams of carbon-12. That number of particles is called Avogadro's number, 6.02×10^{23}.

Nibble. The nibble is a measure of quantity of data or information. It is equal to 4 bits.

Piloerection. The scientific word for when your hair stands on end.

Quad. A quad is a unit of energy equal to 10^{15} British Thermal Units (BTUs). The United States consumes about 100 quads per year, while demand worldwide is 400 quads per year.

Rack unit. One rack unit (U) is 1.75 inches and is used to measure rack-mounted computing and industrial equipment. A 4U computer server case is 7 inches high.

Rad. A rad is a unit of measuring the energy absorbed by a material that has been exposed to radiation. One rad equals 100 ergs per gram of material.

Sagan. A sagan is at least 4 billion of anything. It is a humorous tribute to Carl Sagan (1934-1996) whose catchphrase was "billions and billions."

Schist. A common type of metamorphic rock that can be split easily into sheets.

Scoville heat unit. The Scoville scale is a measure of the hotness of a chili pepper.

Shake. In nuclear engineering and in astrophysics, the shake is a very short period of time. One shake is defined as 10 nanoseconds.

Siriometer. The siriometer is an astronomical measure equal to one million astronomical units, which is one million times the average distance between the Sun and the Earth (93 million miles). The distance is equal to 15.8 light-years, about twice the distance from Earth to the star Sirius.

Sverdrup. One Sverdrup (Sv) is equal to 1 million cubic meters per second, or 264 million gallons per second. It is used in oceanography to measure the volumetric rate of transport of ocean currents.

Oddities of Science

Oddities of Scandals and Frauds

Aggarwal. Dr.Bharat Aggarwal is an Indian American biochemist that worked in cancer research at the University of Texas MD Anderson Cancer Center. In 2012, a fraud was discovered in 65 papers published by him in the area of curcumin (chemical produced in the Curcume longa plants) as a treatment for cancer. Academic whistleblowers turned him in. Aggarwal was reusing data images and manipulated them to represent different results. The journal Biochemical Pharmacology retracted 7 of his articles because "the data integrity has become questionable." In 2015, he retired from the University of Texas MD Anderson Cancer Center. In February 2018, a cancer conference in India was co-organized by Aggarwal. He promoted it with the false claim that it was co-sponsored by MD Anderson Cancer Center. [source: Chu, "Curcumin trial retraction latest in long line for retired researcher," *NutraIngredients.com*, Apr 13, 2018]

Ahimastos. In 2015, Dr. Anna Ahimastos, a high-profile scientist at the Baker IDI Heart and Diabetes in Melbourne, Australia, admitted to fabricating scientific results published in numerous major international medical journals. Several of her articles about a three-year clinical trial involving medications (Ramipril and Prilace) used to treat hypertension were retracted. She later resigned from her position. [source: Scott and Branley, "High-profile researcher admits fabricating scientific results in major journals," *ABC News*, Sep 16, 2015]

Alsabti. Elias Alsabti (1954-1990) was an Iraqi medical researcher who was exposed as a fraud. He plagiarized as many as 60 papers in the field of cancer research, many with non-existent co-authors. He also pretended to be a member of the Jordanian royal family. In 1981, he passed his medical examination in Indiana without have completed any medical school. He died in an auto accident in South Africa in 1990. [source: "Elias A. Alsabti vs. Board of Registration In Medicine," *Justia US Law*, April 10, 1989]

Anderson. Graduate student David E. Anderson, formerly of the University of Oregon, was caught knowingly falsifying research data, voiding result of four studies. Anderson falsified data by removing outlier values or replacing outliers with mean values to produce results that conform to predictions. Anderson and his advisor, Dr. Edward Awh, had 9 of their publications retracted due to data fabrication. Anderson admitted his misconduct. Awh since moved to the University of Chicago. [source: Read, "Feds find UO grad student falsified research data, voiding results of four studies," *The Oregonian*, Jan 9, 2019]

Anversa and Leri. Piero Anversa and Annarosa Leri were two medical researchers affiliated with the Harvard Medical School. In 2011 and 2012, they published research papers on endogenous cardiac stem cells. They claimed that they could regenerate heart muscles by injecting bone marrow cells into damaged hearts. No one else in the world could duplicate their experiments. They later admitted to having "fictitious data points" and "altered figures." They also included "false scientific information" in their grant applications. As a result, two Harvard Medical School hospitals had to pay a $10 million settlement to the US government, and clinical trials based in Anversa and Leri's work was stopped. 31 publications from the Anversa/Leri research group have been retracted. In 2015, Anversa left Harvard Medical School and the Brigham and Women's Hospital after several of his research papers were retracted. [sources: O'Riordan, "Stem Cell Research – Shattered After Fabrication Scandal," *tcdmd.com*, April 12, 2019 and Kolata, "He Promised to Restore Damaged Hearts, Harvard Says His Lab Fabricated Research, *New York Times*, Oct 29, 2018]

Bezwoda. Dr. Werner Bezwoda is a South African clinical investigator who was formerly associated with the University of Witwatersrand. In 1995, he presented his scientific studies for curing late-stage breast cancer to medical conferences and published his results in major medical journals – offering results that seemed too good to be true. And they were. He reported that 51% of his patients achieved complete remission through his high-dose chemotherapy and bone marrow transplant, but his results were fabricated. Bezwoda later admitted to scientific misconduct in trials on high-dose chemotherapy on breast cancer, stating that he had "committed a serious breach of scientific honesty and integrity." He was dismissed from his university in March 2000. Because of Bez-

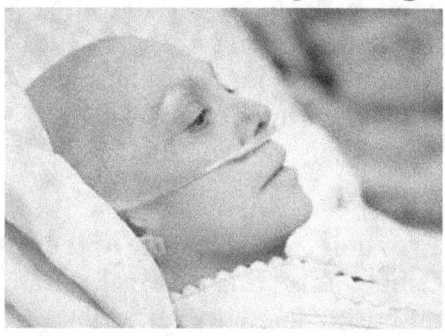

Oddities of Science

woda's fraudulent claims, some 35,000 women used his therapy treatment, and an estimated 9,000 died from the side effects. [source: Gaddy, "The Fraudulent Study That Killed Thousands of Breast Cancer Patients," *ozy.com*, Dec 5, 2016]

Boldt. Joachim Boldt (1954-) is a German anesthesiologist formerly a professor at the University of Giessen. In 2010, he was suspended from Klinikum Ludwigshafen, a hospital in Germany, for a scientific publication with insufficient background research. In 2011, he was stripped of his professorship and criminally investigated for forgery in his research studies. He has had 100 of his publications retracted, the second highest number of scientific retractions ever recorded. 89 of his 102 studies published by Boldt contained research without proper institutional review board approval. [sources: Marcus, "A scientist's fraudulent studies put patients at risk," *Science*, Oct 26, 2018 and Wise, "Boldt: the great pretender," *bmj.com*, Mar 19, 2013]

Bridges. C. David Bridges is a vision researcher at Purdue University, formerly at the Baylor College of Medicine in Houston. He was found by a National Institutes of Health (NIH) investigation to have stolen ideas from a rival's scientific manuscript that Bridges has been asked to review in 1987, and then produced and published those ideas in his own research. The controversy involves around the discovery of an enzyme in the eye. The NIH later stripped Bridges of his funding. [sources: *Chemical Engineering News*, Aug 7, 1989 and Broad, "Question of Scientific Fakery Is Raised in Inquiry," *New York Times*, Aug 12, 1989]

Bulfone-Paus. Dr. Silvia Bulfone-Paus is an Italian immunologist. She had 13 of her publications retracted following investigation of alleged scientific misconduct involving image manipulation of scientific results for several years. In 2011, her laboratory at the University of Manchester was downsized and funding was decreased for the lab. [sources: Marcus and Oransky, "Do scientific fraudsters deserve a second change?" *statnews.com*, June 24, 2016 and "The strange and murky case of Silvia Bulfone-Paus: 12 retractions so far…", The k2p blog (*ktwop-.com*), Jan 25, 2011]

Burt. Sir Cyril Burt (1883-1971) was Britain's most eminent educational psychologist and geneticist. He is known for his studies on the heritability of IQ. He popularized IQ test in the United Kingdom. After he died, it was discovered that he faked the statistics in his IQ studies, and of inventing co-authors. He had fabricated data purporting to show that human intelligence is genetically determined. He also invented correlations in 53 separated twins which did not exist. His studies of inheritance and intelligence have now been discredited. [source: "The Intelligence Fraud," *Socialist Review*, April 1966]

Chandra. Dr. Ranjit Chandra (1938-) is an Indian-born Canadian researcher. He was investigated for suspected research fraud and published several studies on nutrition that have since had their integrity repeatedly questioned. He wrote papers claiming that a mixture of vitamins and minerals could reverse dementia in elderly people. His claims were completely invented. When asked to produce the raw data to back up his claims, Chandra claimed it had been stolen or the university (Memorial University in Newfoundland) had lost it. He was supposedly nominated twice for the Nobel Prize in medicine. But he was exposed as a fraud. When he divorced his wife, he had 120 different bank accounts in various tax havens. He was making money from his patented multivitamin mixture, selling it as an "evidence based" nutritional supplement. The "evidence" came from his dubious clinical trials. He had earlier won the prestigious Order of Canada medal, but his prize was stripped following accusations of his scientific wrongdoing in his research. [source: Hamblin, "The Secret Life of Dr. Chandra," *bmj*, Feb 11, 2006]

Chen. Ching-Shih Chen is the former chair of the cancer research at Ohio State University (OSU). After an anonymous tip, an investigation began and found that he mishandled images and figures in published papers since 2001. He intentionally falsified data and did not keep any laboratory notebooks on his research, in violation of federal research policies. The falsified data was related to anti-cancer small molecule agents that Chen developed. He has had 8 research publications retracted. In September 2017, Chen admitted to research misconduct and resigned his positions at OSU. In 2010, Chen received the Innovator of the Year award at OSU. The university later determined that Chen's awarded research was falsified. [source: "Professor resigns after research misconduct investigation," *osu.edu*, Mar 30, 2018]

Oddities of Science

Cold fusion. Cold fusion is a hypothesized type of nuclear reaction that would occur at, or near, room temperature. In 1989, two electrochemists, Martin Fleishmann and Stanley Pons, reported that their apparatus had produced cold fusion. However, many scientists tried to replicate the experiment, but were unable to do so. There were more than 200 experiments to investigate whether nuclear reactions generating more energy than they consume can occur at room temperature. The Department of Energy concluded that this was impossible to reproduce, and thus probably false. [source: Ritter, "Cold fusion died 25 years ago, but the research lives on," *Chemical & Engineering News*, Nov 7, 2016]

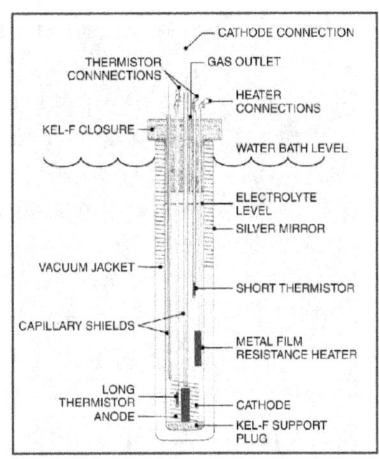

Darsee. John Darsee (1948-) is an American physician and former medical researcher. He was accused of fabricating data in published research articles and more than 100 abstracts and book chapters. He has had at least 17 of his publications retracted. In 1983, he was disbarred for 10 years by the National Institutes of Health (NIH) after their investigation showed that Darsee committed wide-ranging scientific misconduct and fabricating large amounts of data from experiments which he had never conducted. Brigham and Women's Hospital, where Darsee worked, had to return over $122,00 in research funds to NIH. In 1984, New York revoked his license to practice medicine in the state of New York. [source: Broad, "Notorious Darsee Case Shakes Assumptions About Science," *New York Times*, June 14, 1983]

Das. Dr. Dipak Das (1947-2013) was a former director of the Cardiovascular Research Center at the University of Connecticut Health Center. After an anonymous allegation of research irregularities, he was found guilty of 145 counts of fabrication of research data. He had 20 of his publications retracted. In 2012, he was fired from the University of Connecticut Health Center. The Health Center terminated all research in Das's laboratory and declined federal research grants awarded to him. [source: Wade, "Fraud Charges to Dipak K. Das, a University of Connecticut Researcher, *New York Times*, Jan 12, 2012]

Fujii. Dr. Yoshitaka Fujii is a Japanese researcher in anesthesiology and ophthalmology. Since 1993, he has fabricated at least 190 scientific papers, a record for the number of papers by a single author requiring retractions. He was also found to have forged the signatures of scientists has co-authors without their knowledge. In February 2012, after investigations into allegations of scientific mistrust in his studies of the effects of drugs intended to prevent nausea after surgery, he was dismissed from his position as professor of anesthesiology at the Toho University Faculty of Medicine. [source: Stromberg, "Meet Yoshotaka Fujii, the most prolific fraudster in modern science,"*vox-.com*, Mat 21, 2015]

Han. Dr. Dong-Pyou Han is a former researcher at Case Western University and professor of biomedical sciences at Iowa State University. He was working with rabbits to develop an HIV vaccine. In 2013, he added human antibodies to samples of rabbit blood in an effort to make it appear as though a vaccine he was working on had exhibited anti-HIV activity. He was able to convince the NIH to give him up to $20 million in federal grants to work on anti-HIV vaccine. After Han's fraud came to light, he resigned his post in October 2013. In June 2014, as a result of his receiving grant money due to falsified results, he was indicted on 4 federal felony counts of making false statements. In July 2015, he was sentenced to 57 months imprisonment for fabricating data in HIV vaccine trials. [source: Bernstein, "HIV researcher found guilty of research misconduct sentence to prison," *Science*, July 6, 2015]

Hauser. Marc Hauser (1959-) is an evolutionary biologist and former Professor of Psychology at Harvard. In 2010, he was found guilty of scientific conduct. He resigned from Harvard in 2011. It is ironic that he wrote a book about human morality called *Moral Minds* (2006). [sources: *Harvard Magazine*, Sep 5, 2012 and Gross, "Disgrace: On Marc Hauser]

Hwang Woo-suk. Hwang Woo-suk is a South Korean veterinarian and former Professor of Biotechnology at Seoul National University. He committed "deliberate fabrication" in his research on stem cells. Ethical violations included the use of eggs from his graduate students and from the black market. His human cloning experiments were revealed to be fraudulent. In 2006, he was dismissed from his university. In 2008, he was found guilty of embezzlement and bioethical violations in connection to his research program. He received a sentence of 3 years jail time. [source: Cyranoski, "Woo Suk Hwang convicted, but not of fraud," *Nature*, Oct 26, 2009]

Jacobsen. Cecil Jacobsen (1936-) is an American former fertility doctor who used his own sperm to impregnate his patients without informing them. In the 1980s, he operated a reproductive genetics center in Fairfax County, Virginia. He injected his patients with the hormone hCG used as a parenteral fertility medication.  He told his patients that all the standard pregnancy tests were positive and their bodies were beginning to undergo the normal changes. He performed ultrasounds, identifying a fetus in the grainy image. Around the third month, he would report that the fetus had died. In fact, these patients were never pregnant, and the bodily changes were due to the reaction of the hormone. Some patients had arranged to be artificially inseminated, supposedly with sperm provided by screened, anonymous donors. In fact, he was using his own sperm. He has been suspected that he fathered at least 75 children with his own sperm. In 1992, he was convicted of 52 counts of mail fraud, wire fraud, and perjury. He was sentenced to 5 years in prison for his fraud, and had his medical license revoked. [sources: "Cecil Jacobson, The Baby Maker," *Medical Bag*, Mar 25, 2015 and "Doctor is Found Guilty in Fertility Case," *New York Times*, Mar 5, 1992]

Jamal. Dr. Sophie Jamal is a former Professor of Medicine at the University of Toronto and former staff endocrinologist. She falsified data to make her results look better from studies from nitroglycerin compounds in osteoporosis. In 2016, she received a lifetime ban from the Canadian Institutes of Health Research (CHIR). In 2017, she resigned from the university and hospital. In 2018, she had her medical license revoked. [source: Shuchman, "Misconduct saga rattles bone scientists," *CMAJ*. Sep 20, 2016]

Koren. Gideon Koren is an Israeli-Canadian pediatrician. He was a former Directory of the Motherisk Program at the Hospital for Sick Children in Toronto. He was at the center of the Motherisk scandal, In 2015, an independent review found that the drug testing was inadequate and unreliable for use in child protection proceedings. This threw in double the findings of 16,000 child protection cases. The lab had to be shut down amid all the controversy. Earlier, Koren had to be disciplined for writing harassing anonymous letters. In 2019, he relinquished his license to practice medicine in Ontario. [source: Gallant, "Former head of Sick Kids' Motherisk lab gives up medical license amid investigation." *The Star*, Feb 22, 2019]

Macchiarini. Dr. Paolo Macchiarini was an Italian thoracic surgeon who faked research to claim patients with failing tracheas could regenerated functioning one with their own stem cells. Whistleblowers pointed out his scandals. In 2019, he received a 16-month prison sentence for forging documents and abuse of office. He was accused of unethically performing experimental surgeries, even on healthy patients. In 2012, he was arrested in Italy and charged with asking patients for money to pay for his procedures. He was once investigated for manslaughter after some of his patients died. [source: Frellick, "Investigated for Manslaughter, 'Disgrace' Surgeon Get 16 Months," *Medscape*, Dec 3, 2019]

Poehlman. Eric Poehlman (1956-) was a scientist in the field of human obesity and aging. He was the first academic in the United States to be jailed for falsifying data in a grant application. In 2005, he was charged with making false statements on at least 17 federal grant applications. [source: Klintisch, "Poehlman Sentenced to 1 Year in Prison," *Science*, Jun 28, 2006]

Potti. Dr. Anil Potti is a physician and former professor at Duke University who resigned in 2010. His scientific research included false research data regarding the use of genetic analysis. He exaggerated his credentials and also claimed to be a Rhodes scholar. *60 Minutes* did a profile on Potti, describing his case as "one of the biggest medical research frauds ever." The American Cancer Society suspended over $300,000 of grand dollars what were used for Potti's work in cancer treatment.[source: Kaiser, "Potti found guilty of research misconduct," *Science*, Nov 9. 2015]

Reuben. Scott Reuben (1958-) is an American anesthesiologist. He falsified and fabricated clinical trials involving painkiller medication. He never conducted the clinical trials that he wrote about in 21 journal articles. In 2009, he was sentenced to 6 months in federal prison for healthcare fraud.[source: "Fraud Case Rocks Anesthesiology Community," *Anesthesiology News*. Oc6 8,2010]

Sato. Yoshihiro Sao (-2006) was a Japanese bone-health researcher. From 1996 to 2013, he plagiarized work fabricated data, and forged authorships on over 60 scientific papers that had to be retracted. [source: Else, "What universities can learn from one of science's biggest frauds," *Nature*, June 28, 2019]

Sudbo. Dr. Jon Sudbo (1961-) was an oncologist at the University of Oslo. In 2006, he was found to have manipulated and fabricated data in grant applications and 15 of his papers. He published a paper in *The Lancet*, saying it was based on 908 patients in his study on cancer. This was all fabricated. The editor of *The Lancet* described this as the biggest scientific fraud conducted by a single researcher ever. In 2006, Sudbo resigned from his university and his medical license was revoked. Even his doctoral dissertation was condemned as fraudulent. [source: Marris, "Doctor admits Lancet study if fiction, *Nature*, Jan 18, 2006]

Tuberculin scandal. In 1890, German scientist and physician Robert Koch (1843-1910) discovered tuberculin. It is a combination of proteins that are used in the diagnoses of tuberculosis (TB). The original tuberculin discovered by Koch was a glycerin extract of the tubercle bacilli. Koch announced that it was a remedy for TB. However, reductions in deaths due to TB did not meet those expected of the treatment. In Koch's time, about 1 in 10 Germans died of TB. At the 10th International Medical Congress held in 1890 in Berlin, Koch unexpectedly introduced a cure for TB with is discovery of tuberculin. He did not reveal its composition. Koch then attempted to make money from his discovery, which was held against him since he had conducted his research at a public institution with public money. At the time, regulations for testing medicines did not yet exist. Soon, doctors were performing autopsies on the corpses that took tuberculin. It showed that tuberculin did not kill the bacteria, and even activated latent bacteria that caused TB. Koch was forced to reveal his composition of his secret cure. But even he did not precisely know what it contained. Soon, Koch took off for Egypt to get out of the public eye. The tuberculin scandal was a warning about how not to proceed when testing medicine. Koch went on to win the 1905 Nobel Prize in Medicine for his work in bacteriology. [source: Zuger, "The Remedy: a 19th-Century Bid to Cure TB, *New York Times*, June 24, 2014]

Wakefield. In 1998, Dr. Andrew Wakefield (1957-) published a study in *The Lancet*, claiming that his research indicated a connection between autism and the measles-mumps-rubella (MMR) vaccine. This led to many parents refusing the vaccine shot. This led to a number of measles and mumps outbreaks. Wakefield altered facts about the children in his study, and he was also paid off by a lawyer planning to sue the manufacturer of the vaccine. The British General Medical Council found Wakefield guilty of fraud and misconduct. Wakefield lost his job as a gastroenterologist and had his scientific paper retracted. Wakefield's 1998 paper has been described by scientists as "the most damaging medical hoax of the last 100 years." [source: Boseley, "How disgraced anti-vaxxer Andrew Wakefield was embraced by Trump's America, *The Guardian*, July 18, 2018]

Oddities of Science

Oddities in Science Fiction

Antidepressant. In 1931, Aldous Huxley wrote Brave New World. It featured a mood-altering pill called Soma that acted as an antidepressant. Antidepressants were first developed in the 1950s. In 1988, Prozac was approved by the FDA and became the third-most-common-prescription drug in use by 2008.

Atomic bomb. In 1944, the science fiction story *Deadline* by Cleve Cartmill (1908-1964) depicted the atomic bomb before it was invented and used. The story attracted the attention of the FBI by reason of its detailed description of an atomic bomb. The story appeared in the March 1944 issue of *Astounding Science Fiction*. In 1913, H. G. Wells wrote *The World Set Free*. The story mentions a hand grenade of uranium and described that the weapon be dropped from a plane. Wells also predicted automatic sliding doors in 1899.

Cell Phone. The science fiction TV show, *Star Trek*, introduced the wireless communication device in 1966 that has since become the cell phone. In 1973, the first handheld cellular mobile phone was demonstrated by Motorola engineers. The main designer, Martin Cooper, said that he got the idea of creating a wireless communication device from watching Captain Kirk using his communicator. In 1979, the first commercial automated cellular network was launched in Japan. In 1983, the first handheld mobile device for commercial use became available. In 1989, Motorola introduced the "flip" phone similar to the *Star Trek* device. In the 1990s, digital cellular networks appeared.

Communications satellite. In 1945, Arthur C. Clarke wrote an article in the October 1945 issue of *Wireless World*, describing how radio and TV signals could bounce off satellites in geosynchronous orbit for long-distance communication. In 1963, NASA launched Syncom 2, the first geosynchronous communications satellite.

Credit card.. In 1887, Edward Bellamy (1850-1898) wrote *Looking Backward*. The novel was the third largest bestseller of its time, behind *Uncle Tom's Cabin* and *Ben-Hur*. His character falls asleep in 1887 and wakes up in the year 2000. He discovers that everyone uses credit cards (more like debit cards) and that you could just swipe your card to pay for an item and get a receipt for the transaction. You could use the credit card in other countries. The use of credit cards in the USA originated in the 1920s. In 1950, the first universal credit card became available. In 1958, Bank of America issued the first general-purpose credit card that offered a "revolving credit" feature. In 1969, the magnetic strip became standard on credit cards. In 2015, EMV chips were added to credit cards to help protect buyers against fraudulent card transactions. *Looking Backward* also featured Internet-like delivery of goods and warehouse clubs like Sam's Club or BJ's. Bellamy also predicted sermons and music being available in the home through cable "telephone."

Cyberspace. In 1984, William Gibson (1948-) wrote *Neuromancer*. He predicted computer hacking, virtual reality, the world wide web, and cyberspace a decade before the Internet took the form we know today. Gibson coined the world "cyberspace."

Defibrillator. In 1818, Mary Shelley (1797-1851) wrote Frankenstein. Inspired by galvanism (manipulating muscles with electric current), Mary Shelley's Dr. Victor Frankenstein reanimates dead flesh. In 1947, Dr. Claude Beck (1894-1971) performed the first defibrillation. He saved a 14-year-old patient with a 60-Hz jolt to the heart from his homemade defibrillator: two silver paddles wired to an outlet. By the 1950s, the machines were reviving patients in hospitals worldwide.

Home computer. The first home computer, the Altair 8800, was invented by Dr. Ed Roberts (1941-2010) in 1975. He named it after the Altair Solar System in a *Star Trek* episode.

Man on the moon. In the late 1620s, Francis Godwin (1562-1633), a bishop in the Church of England, wrote *The Man in the Moone*. It was published posthumously in 1638 and is considered one of the first works of science fiction. The book was one of the first to support that planets revolved around the sun and that there may be the possibility of extraterrestrials. It was also the first story to depict weightlessness in space. In 1865,

Oddities of Science

Jules Verne wrote *From the Earth to the Moon*, which described a mission to the moon by 3 Americans in a spacecraft. Verne also speculated about a light-propelled spacecraft. That technology exists today in solar cells. Man finally walked on the moon in July 1969.

Picturephones. In 1911, Hugo Gernsback (1884-1967) wrote a novel called *Ralph 124C 41+*, which first appeared in *Modern Electrics* in 1911. The story included a device called the "telephot" that allowed people to have eye contact while speaking across long distances. AT&T first demonstrated consumer video conferencing at the 1964 World's Fair. Gernsback's novel also predicted television and channel surfing, remote-control power transmission, transcontinental air service, solar energy, sound movies, synthetic milk and foods, artificial cloth, voice printing, tape recorders, and spaceflight.

Radio. In 1889, Jules Verne wrote a short story, "In the Year 2889," and predicted that people would one day listen to news instead of just reading the newspaper. The first radio news program was broadcast Aug 31, 1920 in Detroit, Michigan.

Robots. In 1920, Karel Capek (1890-1938) wrote a play called R.U.R. (Rossum's Universal Robots), which introduced the word robot. Capek was nominated for the Nobel Prize in Literature 7 times, but never won. In 1927, the silent-movie *Metropolis* was released. It was the first on-screen depiction of robots. The inventor in the movie crafts a metallic humanoid robot, which is then reskinned to look more like a human.

Smartwatch. In 1962, the cartoon show *The Jetsons* featured a high-tech watch to make phone calls, look at photos, and watch videos. In 1972, the first LED digital watch was developed. In 1983, Seiko developed a watch that had a video output. In 1994, the first wireless smartwatch was developed. Timex developed a watch capable of downloading data from a computer wirelessly. In 1998, the first Linux smartwatch was developed. In 1999, Samsung developed the first smartwatch to make a call.  In 2003, Skype was released. By 2015, Skype had over 300 million estimated active each month. In 2013, Omate was the first company to design a truly independent smartwatch that could make calls, use maps, and take advantage of Android apps. *The Jetsons* also accurately predicted drones and holograms in 1962.

Space Station. In 1869, the novella "The Brick Moon," by Edward Hale (1822-1909) contains the first known depiction of an artificial satellite or space station. The sphere in the story was intended as a navigational aid, but is accidently launched with people aboard. The author even got the idea of needing 4 satellites visible above the horizon for best results, as in modern day GPS, right. "The Brick Moon" was first released serially in 3 parts in *The Atlantic Monthly* in 1869. The first artificial satellite was Sputnik 1, launched on October 4, 1957. The first space station was Salut 1, launched by the USSR in 1971.

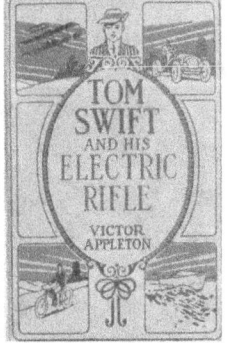

TASER. In 1911, *Tom Swift and His Electric Rifle* was published. It features a hero who invents a gun that fires bolts of lightning. In 1970, Dr. Jack Cover (1920-2009), a NASA physicist, invented a device that could shoot two darts and deliver an electric current designed to disrupt voluntary control of muscles. He named it using an acronym for *Tom A. Swift's Electric Rifle.*

Oddities of Science
Oddities in Myths

Bats are blind. There is a myth that bats are blind. All bats can see in black and white. At night, they see better than we do. They lack color receptors and humans, in dim light, cannot see colors either. Bats also have sonar and can navigate with that. [source: Langley, "6 Bat Myths Busted: Are They Really Blind?"*National Geographic*, Nov 1, 2014]

Birds and humans. There is a myth that a mother bird will reject its baby if the baby bird has been touched by a human. Most birds have a poor sense of smell and wouldn't know the difference if a baby bird was touched or not. Mother birds are much more loyal to their babies than this myth suggests. Birds are quite devoted to their young and not easily deterred from taking care of them. [source: Eveleth, "Do Birds Really Abandon Their Chicks If Humans Touch Them?" *Live Science*, Dec 6, 2011]

Brain. There is a myth that people only used 10% of their brain. You can use a large part of your brain at all times. The brain is 3% of the body's mass, but uses 20% of its energy. [source: Boyd, "Do People Use 10 Percent of Their Brains?" *Scientific American*, Feb 7, 2008]

Bulls and red. There is a myth that bulls become angry at the color red. Bulls and other cattle are partially color-blind and cannot see red. A matador may wave a red cape, but it is the matador's movement of the cape, the taunting and the threats that make a bull charge. The cape is red to mask the bloodstains. [source: "Here's why bullfighter capes Are always red," Business Insider, Sep 27, 2018]

Carrots and night vision. There is a myth that eating carrots helps with night vision. However, eating carrots or other vegetables does not improve your eyesight. Vitamin A (beta carotene) is a major nutrient found in carrots, and is good for the health of your eyes, but will not help those with poor vision. During World War II, sugar was rationed in Britain and carrots on a stick was introduced and advertised as a way to improve night-time vision. [source: Smith, "A WWII Propaganda Campaign Popularized the Myth That That Carrots Help You See in the Dark," Smithsonian Magazine, Aug 13, 2013]

Colds. There is a myth that being cold can give you a cold. There is no evidence that going outside in the cold with wet hair or no jacket will make you sick, provided you don't come down with frostbite or hypothermia. You catch more colds in the winter because you spend more time in close quarters indoors where it is more likely that you will catch the cold-causing virus spread from another person. [source: Fecher, "Do You Really Get Sick from Being Cold?" *Unitypoint.org*, Dec 5, 2017]

Diamonds and coal. There is a myth that diamonds come from compressed coal. Coal is found about 2 miles down from the surface of the Earth. Diamonds are from carbon that is compressed and heated 90 miles down below the surface of the Earth.

Eating and swimming. There is a myth that you need to wait 30 minutes or an hour after eating to swim or you might cramp up and drown. There is no evidence to support this claim. The myth is that digesting food will draw blood to your stomach, meaning less blood is available for your muscles, making them more likely to cramp. There has never been a documented case of anyone drowning because they had a cramp related to swimming with a full stomach. If you get a cramp (not due to your stomach), the best policy is to float for a minute and let it pass. [source: Roth, "Should you wait 30 minutes to swim after eating?" *mayoclinic.org*, Jun 22, 2018]

Elephants and mice. There is a myth that elephants are afraid of mice because they nibble on their feet or can climb up into their trunks. This myth originated from the ancient Greeks and modern children's stories. Elephants have no particular fear of mice. Elephants, however, have poor eyesight and are slow moving. What may happen is that an elephant can be startled when a small animal like a mouse or  bird darts past and surprises the elephant. An elephant, with its trunk, can lift 500 pounds. The trunk is also dexterous and sensitive enough to pick up a mouse, or single blade of grass, or take a coin from the ground. [source: "Are Elephants Really Afraid of Mice?" *Live Science*, Jun 1, 2016]

Oddities of Science

Flat Earth. There is a myth that people in the Middle Ages (and some people today like "Mad" Mike Hughes) thought the Earth was flat. However, during the Middle Ages, almost every scholar thought the Earth was round, not flat. The earliest documentation of a spherical Earth comes from the Ancient Greeks in the 5th century BC. Since the 600s AD, scholars have supported that view. By the Early Middle Ages (700 to 1500 AD), virtually all scholars maintained the spherical viewpoint. [source: Main, "Even in the Middle Ages, People Didn't Think the Earth Was Flat," *Newsweek*, Jan 28, 2016]

Frogs and warts. There is a myth that people get warts from frogs and toads (all toads are frogs). Frogs or toads cannot give you warts, but shaking hands with someone who has warts can give you warts. The human papilloma virus (HPV) is what gives people warts, and it is unique to humans. [source: "Do toads actually give you warts?" *psu.edu* (Penn State University), Sep 16, 2015]

Full moon. There is a myth that the full moon affects behavior. The full moon (or any other phase of the moon) does not draw out strange behavior in people. This myth was supposedly caused by water in the brain, affected by tidal forces of the moon. There is no correlation between the full moon and increased erratic behavior. [source: Golembiewski, "Why Do We Still Believe in 'Lunacy' During a Full Moon, *Discovery Magazine*, Aug 16, 2019]

Glass, a slow-flowing liquid. Glass is not an ultra-slow flowing liquid. This myth arises from the fact that windows glass in older building is typically thicker at the base. Supposedly, the glass was evenly thick on all sides, but gradually flowed downwards over time, thickening the base. In reality, glass is an amorphous solid. In earlier years, it was difficult to get a perfectly even pane of glass using old-fashioned methods. When a craftsman had a flat pieces of glass, he would cut the glass to size. If one side was thicker than the other, he would put that side at the bottom of the pane for stability. [source: Curtin, "Fact or Fiction?: Glass Is a (Supercooled) Liquid," *Scientific American*, Feb 22, 2007]

Goldfish and memory. There is a myth that goldfish have a 3 to 7 second memory. However, goldfish can learn basic survival skills and remember them for up to 3 months, Goldfish can also tell time. In lab experiments, goldfish were conditioned to push a lever for food at roughly the same time each day. [sources: Simpson, "Fish's memories last for months, say scientists, *The Telegraph* (UK), Jan 7, 2009 and "Goldfish Memory: Is 3 second goldfish memory a myth?" thegoldfishtank.com, Dec 1, 2019]

Hair and nails. There is a myth that hair and nails keep growing after death. That is not true. Instead, the skin dehydrates or dries out and shrinks, giving the appearance of further growth. Fingernails grow 0.1 millimeters a day. But in order to grow, they need glucose. Once the body dies, there is no more glucose. [source: Geggel, "Do Hair and Nails Keep Growing After a Person Dies?" *Live Science*, Dec 15, 2016]

Housefly lifespan. There is a myth that houseflies (Musca domestica) have a lifespan of 24 hours. The common housefly has a lifespan of 15 to 25 days. Some specimens live a little longer, perhaps up to 2 months. The mayfly, however, has a lifespan between 5 minutes and 24 hours. [source: Mancini, "Do Flies Really Have 24-Hour Lifespans?" *mentalfloss.com*, Mar 12, 2014]

Lightning. There is a myth that lightning never strikes the same place twice. In reality, lightning often strikes the same place twice, or more. Lightning is the number one cause of storm-related deaths. Lightning globally kills over 6,000 people a year. When we see a lightning strike, we're witnessing the discharge of electricity that has built up in a cloud, which is so strong that it breaks through the ionized air. This creates a lightning bolt that travels downward until it reaches the ground. It is a very

quick process that takes about 30 milliseconds. And right after lightning strikes, it reverberates in quick succession. So, multiple strikes can happen at the same place in this short period of time. Even during the same thunderstorm, there is nothing stopping a lightning bolt from striking the same place it had struck previously, even if it was as little as a few seconds earlier or as much as centuries later. Lightning usually hits the Empire State Building 100 times a year. It was once

Oddities of Science

hit 15 times in 15 minutes. A big flash releases almost 4 billion kilowatts of energy. [source: Childs, "Scientists Have Always Known Lightning Does Strike Twice in the Same Place, But Now They Think They Know Way," weather.com, April 18, 2019]

Ostrich. There is a myth that ostriches stick their heads in the ground when scared. That's just not true. When an ostrich gets scared, it either runs or plays dead. An ostrich can run about 31 mph, faster than any human. They are fast enough to escape most predators. Ostriches do dig a shallow hole in the ground and make their nests there. [source: D'Mello, "Ostrich Head in Sand: Do Ostriches Really Bury Their Heads?" *ScienceABC*, Sep 15, 2015]

Penny dropped. . A science myth is that if you drop a penny (which weighs about 1/11 of an ounce) from the top of the Empire State Building, it would accelerate to the point that it would kill a person on the sidewalk below if it hit that person. In reality, the penny will reach a terminal velocity of about 30 to 50 mph in freefall, depending on the wind, not enough to kill anyone.

Photographic memory. There is a myth that people have photographic memories. There is no such thing as a "photographic" memory. Some people have very good memories, but the human memory cannot recall events with visual details precise enough to mimic the fidelity of film or a camera sensor. A true photographic memory has never been proved to exist. [source: Gordon, "Does Photographic Memory Exist?" *Scientific American*, Jan 1, 2013]

Senses. The myth is that we only have 5 senses – sight, sound, touch, taste, and smell. But we actually have more senses. The others are proprioception – sensing one's orientation in space; thermoception – sense of temperature; equilibrioception – sense of balance; and interoception – sense of one's physiological condition. There is also nociception – the sense of pain, and the sense of time. [source: Perry, "Think you have only 5 senses? You've actually got about 14-20," *bigthink.com*, May 2, 2018]

Sharks and blood. There is a myth that sharks can smell a blood from miles away. Sharks do have a good sense of smell and can detect 1 part blood per 10 billion parts water. That is roughly a drop of blood in an Olympic-size swimming pool. For an ocean, it takes awhile for odor molecules to drift. A shark, sooner or later, will smell blood as far away as **600 feet, but not miles.** [source: "Can sharks smell blood in the water?" *SurferToday*, Jul 31, 2019]

Stress and blood pressure. There is a myth that being stressed will give you high blood pressure (130/80 or higher). However, stress does not play a large role in chronic high blood pressure. Acute stress can temporarily increase blood pressure, but overall, it is not a main cause of hypertension. Hypertension is more likely caused by genetics, bad diet, smoking, and other bigger factors. [source: Aungst, "Anxiety and High Blood Pressure: What's the Connection," *goodrx.com*, Jan 3, 2019]

Sugar. There is a myth that sugar makes children hyperactive. There is no evidence of this. The ramped-up energy in kids is more likely caused by the excitement over getting a treat or being around other kids. However, sugar is associated with weight gain, hypertension, and insulin resistance that may cause diabetes. [source: Ansel, "Sugar: Does it Really Cause Hyperactivity," *eatright.org*, Aug 20, 2018]

Toilet flushes. There is a myth that toilet flushes spin in a different direction in the Southern Hemisphere than the Northern Hemisphere. However, both directions can be found in both hemispheres. This myth is based on the Coriolis Effect, where flow patterns are affected by the earth's rotation. This  explains the spin direction of the Gulf Stream or hurricanes, but not toilet flushes or tornadoes. These are too small to be influence directly by the Coriolis Effect. The directions of a toilet flush is due to the design of the toilet, the plumbing, and water pressure. [source: Feltman, "Myth busted: Water does swirl in different directions across the globe, but not a toilet thing," *Washington Post*, June 3, 2015]

Water and 8 glasses a day. There is a myth that everyone should drink 8 glasses of water a day. This idea of 8 glasses (i.e., 8-oz cups) a day is an odd one. In healthy people, researchers have not found any connection between fluid intake and heart disease, kidney disease, skin quality, or sodium levels. In fact, drinking too much water can be a problem. Just drink when you are thirsty and don't count the glasses. [source: Murphy, "8 glasses a water a day: Myth or medicine?" *MDLinx*, Oct 10, 2018]

Oddities of Science

Oddities of Scientists

Eccentric Scientists

Buckland. William Buckland (1784-1856) was a geologist, paleontologist, and priest. He was President of the Royal Geographical Society. He was the first man to publish the first scientific study of a complete dinosaur skeleton, which he named Megalosaurus. He had the odd habit of insisting on dining of everything. His ambition was to eat an example of every animal in existence. He was known to have eaten roast hedgehog, potted ostrich, panthers, porpoises, puppies, bluebottle flies, crocodile, mice on toast, sea slugs, and garden moles. He drank bat urine and reportedly ate the shrunken heart of King Louis XIV. He said the boa constrictor tasted like veal and that earwigs were very bitter. He allowed a number of wild animals to roam through his house and garden. Charles Darwin thought Buckland was a 'buffoon," (his exact words).   [source: Lewey, "The Man Who Ate Everything," *The Guardian*, Feb 25, 2008]

Dirac. Paul Dirac (1902-1984) was an English theoretical physicist. In 1933, he shared the Nobel Prize in Physics with Erwin Schrödinger (1887-1961) for the discovery of new productive forms of atomic theory and their contributions to quantum mechanics. And at the time, Dirac was the youngest person ever to receive that honor. He was known to exhibit autistic traits that made him appear very unusual. Dirac was regarded by his friends and colleagues as unusual in character. His colleagues in Cambridge jokingly defined a unit called a "dirac," which was one word per hour. Dirac relaxed by climbing trees in a three-piece suit. When his wife once got mad and exploded, "What would you do if I left you?" he thought for a bit then replied, "I'd say, 'Goodbye, dear." [sources: "Paul Dirac: The man who conjured laws of nature from pure thought, *The Guardian*, April 2, 2009 and Farmelo, *The Strangest Man,* 2009]

Flammarion. Nicolas Camille Flammarion (1842-1925) was a French astronomer and author of over 50 scientific titles. He believed that people lived on Mars and they tried to communicate with the Earth in the past. He scared everyone when he believed that the tail of Halley's comet had gas that would wipe out all life on the planet. He believed in ghosts, telepathy, and life after death. He accepted spirit communications and he said he was in intimate relations with spirits who had already lived on Earth. [source: "Nicholas Flammarion," *Psi Encyclopedia*, Aug 26, 2019]

Einstein. Albert Einstein (1879-1956) was a German-born theoretical physicist who developed the theory of relativity. He was mildly eccentric. In 1901, he wrote his first scientific paper on the physics of fluids in drinking straws. His second marriage was to his cousin, Elsa. He divorced twice on unfriendly terms. He had multiple affairs. He never wore socks and bragged about it. He smoked cigarette butts picked up off the streets and used any excess tobacco for his pipe. He slept more than 10 hours a night. He also took regular naps. He was a hardened pipe smoker. He ate grasshoppers to gross out his friends. He walked around his neighborhood playing the violin while watching birds. He rarely and reluctantly took baths or showers. Einstein himself never saw the need, so it was up to close friends to practically drag him into a shower. He loved to sail, but never learned to swim and had to be saved one time when his boat capsized on a lake. He finally died after refusing an easy operation for internal bleeding. He had been suffering from an abdominal aortic aneurism for several months. [source: Isaacson, *Einstein: His Life and Universe*, 2008]

Heaviside. Oliver Heaviside (1850-1925) was a self-taught mathematical physicist and electrical engineer. He predicted the existence of the ionosphere, an electrically conductive layer in the upper atmosphere (the Heaviside Layer) that reflects radio waves. He patented the first coaxial cable. He was a genius, but unemployed for most of his career. In his later years, he became an eccentric recluse. He painted his fingernails bright pink (perhaps to cover up the yellowing due to jaundice) and had granite blocks moved into his house for furniture. He claimed he only consumed milk for days, a habit that Thomas Edison and Nikola Tesla had. He refused to let friends and neighbors step inside his house. His closest scientific friend described him as "a first-rate oddity." [source: Long and Wilson, *The Mad Scientist Hall of Fame*, 2008]

Oddities of Science

Newton. Isaac Newton (1642-1727) was one of the most influential scientists of all time. His book, *Principia Mathematica*, formulated the laws of motion and universal gravitation. He was the second scientist  to be knighted, after Sir Francis Bacon. Newton had his share of eccentricities. He had a low esteem of himself and had no close relationships. He distrusted his friends and was suspicious of others. In more than 30 years as a Fellow of Trinity College, he tutored only 3 students, none of whom proceeded as far as a bachelor's degree. He was reluctant to publish his work on mathematics and calculus because he feared controversy and criticism. He spent half his life muddling with alchemy and the occult. He may have been suffering from mercury poisoning from his alchemical experiments. In 1679 he had a nervous breakdown. He was deemed mentally ill again in 1692. His mental illness lasted 18 months. During that period, he broke all contact with his friends and colleagues, crawled into corners, accused everyone of plotting against him, and reported conversations that never took place. He died in his sleep in London on March 31, 1727. His body was buried in Westminster Abbey. [source: "Isaac Newton Biography," *Live Science*, March 24, 2016]

Oppenheimer. Julius Robert Oppenheimer (1904-1967) was an American theoretical physicist and is sometimes called the "father of the atomic bomb." In 1924, he left a poisoned apple for his tutor, Patrick Blackett (1897-1974) while at Cambridge because he didn't like him. The apple was doused with noxious chemicals on British experimental physicist Blackett's desk. Blackett did not eat the apple. Oppenheimer's father had to lobby the authorities to prevent Robert from being charged with attempted murder or other criminal charges. He was 6 foot, 2 inches tall and never weighed over 130 pounds throughout his life. Oppenheimer told his brother that he needed physics more than friends. [source: Veisdal, "The Eccentric and Ingenious Father of the Atomic Bomb. J. Robert Oppenheimer," *medium.com*, Dec 26, 2019]

Tesla. Nikola Tesla (1856-1943) was a Serbian-American inventor, electrical engineer, and mechanical engineer. He was the major contributor in the design of alternating current (AC) electricity. In 1909,

Marconi received the Nobel Prize in Physics for the development of the wireless radio. In 1915, Tesla unsuccessfully sued Marconi, claiming infringement on his patents. Tesla was also very eccentric. He hated jewelry (especially pearls) and round objects. He would not shake hands with anyone over concern with germs. He feared germs and practiced very strict hygiene. He wore gloves to avoid any physical contact with people he met. He was obsessed with feeding pigeons. He used 18 napkins every meal and required 18 towels every morning. He hated overweight people (Tesla was 6 feet 2 inches and weighed 142 pounds). He fired his secretary because of her weight. Besides walking, he exercised by curling his toes 100 times each foot every night, saying that it simulated his brain cells. He had a debilitating form of Obsessive-Compulsive Disorder (OCD). He insisted all his food be boiled before he would touch it. He would walk three times around a building before entering it. He claimed that his celibacy played an important role in his creativity. He died alone, in poverty, and relative obscurity in 1943. [source: Gunderman, "The Extraordinary Life of Nikola Tesla," *Smithsonian Magazine*, Jan 5, 2018]

Murdered Scientists

Charlois. Auguste Charlois (1864-1910) was a French astronomer who discovered 99 asteroids during his life. On March 26, 1910, he was murdered by Gabriel Brengues, the brother of his first wife, Jeanne. They fought over an inheritance issue after the death of Charlois's first wife. Brengues was given a life sentence of hard labor in New Caledonia.

Oddities of Science

Drummond. Sir Jack Drummond (1891-1952) was a distinguished British biochemist, noted for his work on nutrition as applied to the British diet under rationing during World War II. He was the first person to isolate pure Vitamin A. During World War II, he was chief advisor on food contamination. At the time, Drummond was Britain's top food chemist. He was beaten and shot in France while on vacation, along with his wife and 10-year-old daughter. They were camping near the village of Lurs,  on the banks of the Durance River. Gaston Dominici was convicted of the 3 murders and sentenced to die by guillotine. The sentence was later commuted to life imprisonment. Dominici was never pardoned or given a re-trial and died on April 4, 1965. [source: Eddy, "The Mysterious Murder of a Heroic Chemist Who Was (Probably) Also a Spy," *Gizmodo*, Aug 20, 2015]

Dyck. Dr. Leonard Dyck (1955- 2019) was a professor of botany at the University of British Columbia. He was murdered in northern British Columbia by a couple of teenage killers, Kam McLeod and Bryer Schmeglsky. [source: Williams, "Plant Biologist Killed on Solo Camping Trip in British Columbia," The Scientist, Aug 8, 2019]

Eaton. Suzanne Eaton (1959-2019) was an American molecular biologist. She was raped and murdered on July 2, 2019 while attending a conference in Crete. 27-year-old Giannis Paraskasis admitted to deliberately hitting her with his car, raping her, choked her to death, and dumped her body in a ventilation drain of an abandoned World War II bunker outside the port city of Chania. An autopsy showed that she also had broken bones in her face and ribs, and injuries to her hands. [source: "Suzanne Eaton case: Man confesses to killing U.S. scientist," *Washington Post*, July 16, 2019]

El al-Mashad. Yahya El al-Mashad (1932-1980) was an Egyptian nuclear scientist who headed up the Iraqi nuclear program. He was killed on June 14, 1980, in a Paris hotel room in an operation attributed to the Israeli Mossad. Israel denied involvement. Some sources state that Mashad's throat was cut and he had multiple stab wounds. Other sources say that he was bludgeoned to death. [source: Peck,

"The Long Tradition of Killing Middle Eastern Nuclear Scientists, *The Atlantic*, Jan 12, 2012]

Fossey. Dian Fossey (1932-1985) was one of the foremost primatologists in the world. She was a world leading expert on mountain gorillas. In December 1985, she was discovered murdered, killed by a machete blow to the face. To this day, her murder remains unsolved, probably killed by poachers. [source: Hogenboom, "The woman who gave her life to save the gorillas," BBC, Dec 26, 2015]

Gibbins. Ernest Gerald Gibbins (1900-1942) was a British entomologist who discovered 26 new species of insects. In November 1942, while researching tropical diseases in Uganda, he was speared to death by tribesman who thought that Gibbins was using their blood samples for "white man's witchcraft." His car had been ambushed by Lugbara tribesman as Gibbons was returning to his home in Entebbe, Uganda. [source: Flowers, "Five high profile conservationists who were murdered trying to protect the planet," *Blasting News,* Nov 19, 2015]

Hosseinpour. Ardeshir Hosseinpour (1962-2007) was an Iranian nuclear scientist. On January 15, 2007, he was found dead at his home. The Fars News Agency said that he suffocated by fumes from a faulty gas line while he slept. A private intelligence company released a report that he was killed by radioactive poisoning. He may have been murdered by the Israeli Mossad or by Iran's Revolutionary Guards. His sister says he was killed after refusing to help build a nuclear weapon. [source: Ho, "Iran accused of assassinating its own nuclear scientist," *The Times of Israel*, Sep 30, 2014]

Katzir. Aharon Katzir (1914-1972) was a chemist and a pioneer in the study of electrochemistry of biopolymers. He was murdered in a 3-member Japanese (Japanese Red Army) terrorist attack at Lod (later Ben Gurion International Airport), Israel on May 30, 1972. Twenty-six people were killed and 78 were injured.

Kumar. Suresh Kumar (1961-2019) was a scientist for the Indian Space Research Organization (ISRO). He worked in the photography department of India's National Remote Sensing Center. He was found murdered in his house in Hyderabad in September 2019. He was hit on the head with a heavy object, resulting in his death. [source: *The News Minute*, Oct 3, 2019]

Oddities of Science

Marks. Dr. Rodney Marks (1968-2000) was an Australian astrophysicist. He died mysteriously on May 12, 2000, at a South Pole research station while working on the Smithsonian Astrophysical Observatory. His body was held by the government and security contractor Raytheon for unknown reasons. It was later revealed that Marks died of 150 mL of methanol (methyl alcohol) poisoning and died of heart failure. The case received media attention as the "first South Pole murder." The cause of the fatal methanol poisoning has never been determined. [source: "Rodney Marks," *Unresolved*, Nov 3, 2019]

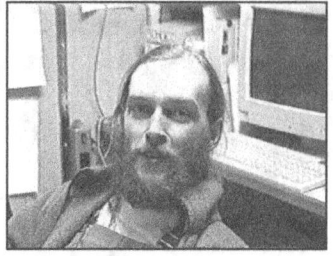

Mouzoko. In April 2019, Dr, Richard Mouzoko (1977- 2019) was shot dead at a hospital in Butembo, Democratic Republic of Congo, where he was treating Ebola patients. Three Congolese doctors were arrested over the killing. Over 200 health facilities have been attacked in the Congo, forcing health workers to suspend or delay Ebola vaccinations and treatments. [source: "DR Congo medics arrested over death of Ebola doctor," *BBC News*, Aug 8, 2019]

Numerov. Boris Numerov (1891-1941) was a Russian astronomer and geophysicist. He was the founder and director of the Institute for Theoretical Astronomy in Leningrad. In October 1936, he was arrested and sentenced to 10 years hard labor. He was accused of being a spy for the Germans. The basis of this odd accusation rested on the fact that German astronomers named an asteroid after him. He was executed by firing squad, along with other political prisoners, on September 13, 1941, but not before supposedly implicating almost the entire staff of the Pulkovo Observatory as fascist spies. This led to their arrests and Pulkovo's demise as one of the world's greatest observatories.

Pollock. Dr. Steven Pollock (1947-1981) was a mycologist (study of fungi or mushrooms) who studied mushrooms and published many articles on the potential of mushrooms to treat illness. On February 1,

1981, he was found shot dead in the forehead at his home in San Antonio, Texas. There were three suspects, but no one was arrested or went to prison for the crime. All three suspects died free men. [source: "Mushroom Madness," *NPR*, Jan 30, 2015]

Roshan. Ahmadi Roshan (1979-2012) was an Iranian nuclear scientist. He was assassinated on January 11, 2012 by a motorbike bomb attack in Tehran. [source: Johnson, "Mossad Agents Assassinated the Iranian Scientist Last Week – Sunday Times," *Business Insider,* Jan 12, 2012]

Sambelashvili. Dr. Aleksandre Sambelashvili (1977-2019) was a biomedical research scientist at Medtronic in Minneapolis. He worked on implanted CRT devices. In August 2019. Erik Kravchuk was charged with first-degree manslaughter after head-butting Sambelashvili at a downtown Minneapolis bar. Sambelashvili died of a traumatic brain injury after falling backward to the floor and hitting his head. [source: Austin, "Man charged with manslaughter in assault which killed Medtronic scientist." *kare11.com*, Sep 13, 2019]

Sartory. Walter Sartory (1936-2003) was a retired nuclear physicist and mathematician at the Oak Ridge National Laboratory in Tennessee. In 2009, he was kidnapped, murdered, and his body was burnt by Willa Blanc (1961-), a cleaning lady. She had help from her son, Louis Wilkinson. After Sartory's death, Blanc used forged documents to raid his bank accounts, getting more than $200,000. She was sentenced to life without parole. [source: Donaldson, "Death and the Maid," *Cincinnati Magazine*, June 10, 2013]

Schlick. Moritz Schlick(1882-1936) was a German physicist. He published papers on relativity. He was the founder of a group of scientists and philosophers known as the Vienna Circle. On June 22, 1936, while ascending the steps at the University of Vienna for a class, he was confronted by a former student, who killed  Schlock with a pistol. The assassin, Johann Nelbroeck (1903-1954), was released on probation after serving two years of a 10-year sentence.

Schwartz. Robert Schwartz (1944-2001) was a nationally renowned scientist in the field of biometrics and DNA research. On

Oddities of Science

December 8, 2001, Schwartz was stabbed to death by Kyle Hulbert at his farmhouse in Leesburg, Virginia. Kyle was directed by Clara Swartz, Robert's daughter, to kill her father, as part of a role-playing game called Underworld. Hulbert was sentenced to life imprisonment and Clara was sentenced to serve 48 years in prison. [source: "Daughter Guilty in Dad's Sword Killing," *Los Angeles Times*, Oct 16, 2002]

Shahriari. Majid Shahriari (1966-2010) was a nuclear scientist and engineer who worked with the Atomic Energy Organization of Iran. On November 20, 2010, assassins riding motorcycles planted a bomb on his car windows whilst he was driving to work in Tehran. The bomb detonated, killing Shahriari instantly. Blame was put on Israel and the United States for the assassination. [source: "Iranian nuclear scientist killed in motorbike attack," *BBC News*, Nov 29, 2010]

Vavilov. Nikolai Ivanovich Vavilov (1887-1943) was a prominent Soviet agronomist, botanist, and geneticist. He was considered the Charles Darwin of the plant world. He was the first to identify the centers of origin of cultivated plants. In 1941, his work was criticized by biologist Trofim Lysenko (1898-1976), who was a favorite with Joseph Stalin. As a result, Vavilov was arrested in 1941, charged with sabotaging Soviet agriculture, and sentenced to death. His sentence was later commuted to 20 years in prison. He was starved to death in prison. In 1948, the study of genetics was completely banned in the Soviet Union. [source: Pringle, *The Murder of Nikolai Vavilov*, 2011]

Webster. David Webster (1944-1989) was an anthropologist in Johannesburg, South Africa. On May 1, 1989, he was shot dead outside his house. A hit squad was paid $8,000 by an apartheid organization for his murder. The triggerman, Ferdi Barnard, that fired the shotgun was caught and later sentenced to two life terms. Barnard was paroled in 2019. [source: "David Webster's killer Ferdi Barnard a 'free man' after two decades," *Sunday Times* (South Africa), Apr 2, 2019]

White. Mary E. White (1926-2018) was an Australian paleobotanist. She established a collection of 12,000 specimens of plant fossils for the Australian Museum in Sydney. On August 5, 2018, she was found dead in her room at an aged care complex in Victoria, Australia. Her 68-year-old daughter was charged with her murder. [source: "Scientist Mary White's daughter, charged with her murder at aged-care facility, granted bail," *ABC News*, Aug 10, 2018]

Zeng Shuo. In 2019, Chinese-born research scientist and engineer Zeng Shuo (1985-2019) was killed on his 34th birthday chasing a laptop thief at a Starbucks in Oakland, California. One man snatched his laptop and ran to a waiting vehicle. Zeng chased after the thief. The car drove off as Zeng grabbed the door handle, causing him to slam his head into a parked car. Bryon Reed and Javon Lee were charged in connection with the theft and special circumstances murder and involuntary manslaughter. Zeng was a research scientist for Aspera, and IBM company. [source: McCarthy, "Pair charged with murder, manslaughter after IBM Aspera boffin killed in New Year's Eve laptop theft struggle," *The Register*, Jan 6, 2020]

Death of Scientists

Aaronson. Marc Aaronson (1950-1987) was an American astronomer. He concentrated his work on the determination of the Hubble constant (giving the age and size of the universe) and the study of carbon-rich stars. He was one of the first astronomers to attempt to image dark matter using infrared imaging. He imaged infrared halos of unknown matter around galaxies that could be dark matter. Aaronson was killed on April 30, 1987 in the dome of one of the observatories at Kitt Peak National Observatory in Tucson, Arizona. He was crushed to death by the hatch leading out to the catwalk when the hatch was slammed shut on him as the 150-ton revolving telescope dome revolved. [source: *Los Angeles Times*, May 2, 1987]

Bogdanov. Alexander Bogdanov (1873-1928) was a Soviet physician, science fiction writer, and pioneer of cybernetics. He was a pioneer in hematology and was founder of the Institute for Hematology

Oddities of Science

and Blood Transfusions In 1924, he started blood transfusion experiments looking for a way to achieve eternal youth or at least partial rejuvenation. He underwent 11 blood transfusions himself, stating that it cured his baldness and improved his eyesight. In 1928, he took the blood of a student suffering from malaria and tuberculosis. This led shortly to Bogdanov's death, but the student with Bogdanov's blood made a full recovery. [source: Inglis-Arkell, "The Genius Who Killed Himself Trying to Become Immortal," *Gizmodo*, March 25, 2014]

Brahe. Tycho Brahe (1546-1601) was a well-known Danish astronomer who made the most accurate celestial observations of his

time. He lost his nose in a sword duel, and had an artificial nose made of brass. He probably died of a burst bladder after attending a banquet in Prague and failing to relieve himself after several hours due to proper etiquette. A later investigation thought he died of mercury poisoning. It was speculated that he had been intentionally poisoned by either his assistant or his cousin. His body was exhumed twice, in 1901 and 2010. The latest conclusion was that he likely died of a bladder infection. Traces of mercury were found in his body, but not enough to kill him. [source: Redd, "Tycho Brahe Biography," *space.com*, Sep 13, 2017]

Buchner. Dr. Eduard Buchner (1860-1917) was a German chemist. In 1907, he was awarded the Nobel Prize in Chemistry for his work on fermentation. He demonstrated that fermentation could be done using chemicals rather than living yeast. At the outbreak of World War I, he volunteered and rose to the rank of Major, commanding a munitions-transport unit. On August 11, 1917, while stationed at Focsani, Romania, he was hit be a shell fragment and died on August 13, 1917. He died during the Battle of Marasesti, the last major battle between the German Empire and the Kingdom of Romania. German casualties were around 60,000 men, while Romanian casualties amounted to 27,000 men. Buchner is buried in the cemetery of German soldiers from Focsani, Romania. [source: "Eduard Buchner," *Famous Scientists*, April 15, 2017]

Curie, Pierre. Pierre Curie (1859-1906) was a French physicist who received the 1903 Nobel Prize in Physics with his wife. On April

19, 1906, while crossing the busy Rue Dauphine in the rain in Paris, he slipped and fell under a heave horse-drawn carriage. He died when one of the wheels ran over his head, fracturing his skull. He was only 46. Had Pierre Curie not been killed, he would have died of the effects of radiation, like his wife and daughter. [source: "Pierre Curie," *biography.com*, July 26, 2019]

Daghlian. Harry Daghlian (1921-1945) was an American physicist who worked on the Manhattan Project to design the first atomic bomb. On August 21, 1945, during a critical mass experiment, he accidently dropped the last tungsten carbide brick being used as a neutron reflector into a plutonium bomb core. The brick caused the nuclear reaction to go critical. He tried to knock the brick away, but had to remove the bricks by hand to halt the reaction. He stopped the reaction, but was exposed to massive amounts of radiation. He died 25 days later. He was the first known fatality cause by a criticality accident. [source: Dockrill, "The Chilling Story of WWII's 'Demon Core' and the Scientists Who Became Its Victims," *Science Alert*, May 5, 2019]

Diesel. Rudolf Diesel (1858-1913) was a mechanical engineer and German inventor. He invented the Diesel engine in 1897. In September 1913, Diesel boarded the German steamer SS Dresden in Antwerp on his way to meeting the Diesel Manufacturing company in London. After the first day, he was never seen alive again. Ten days later, his body was found floating in the North Sea near Norway. He may have committed suicide or was murdered. [source: Latson, "The Mysterious Disappearance of the Diesel Engine's Inventor, Time, Sep 29, 2015]

Drew. Dr. Charles R. Drew (1904-1950) was an African-American surgeon. He researched in the field of blood transfusions. He pioneered the methods of storing blood plasma and organized the first large-scale blood bank. He directed the blood plasma programs of the United States and Great Britain during World War II, but resigned after a court ruling that the blood of African-Americans would be segregated. On April 1, 1950, he was driving his car with 3 other black physicians from Tuskegee, Alabama back to Washington, D.C. when he lost control of his car in North Carolina. The car flipped 3 times. The three other physicians suffered minor injuries, but Drew was trapped in the car and died in a hospital in Burlington, North Carolina. [source: "Charles Richard Drew," *Science History Institute*, Dec 4, 2017]

Fleischman. Elizabeth Fleischman-Ascheim (1867-1905) was an 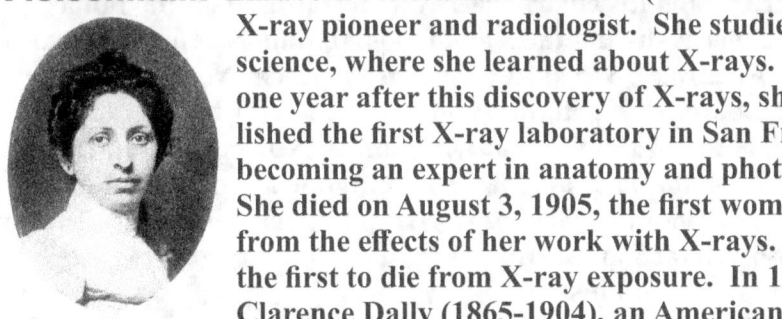 X-ray pioneer and radiologist. She studied electrical science, where she learned about X-rays. In 1897, one year after this discovery of X-rays, she established the first X-ray laboratory in San Francisco, becoming an expert in anatomy and photography. She died on August 3, 1905, the first woman to die from the effects of her work with X-rays. She wasn't the first to die from X-ray exposure. In 1904, Clarence Dally (1865-1904), an American glassblower and assistant to Thomas Edison in his work on X-rays, died under similar circumstances to Fleischmann. Following this, Edison abandoned his research on X-rays. [source: Abramovich, "Elizabeth Fleischmann-Aschheim," *Hektoen International*, Jan 30, 2017]

Gascoigne. William Gascoigne (1612-1644) was an English astronomer. He invented the micrometer and the telescopic sight. His telescopic sight invention came to him when a thread from a spider's web happened to become caught at exactly the combined optical focal points of two lenses. He realized that he could more accurately point the telescope using the line as a guide, and went on to invent the telescopic sight by placing crossed wires at the focal point to define the center of the field of view. In 1642, civil war broke out in England. Gascoigne received a commission in the army of King Charles I. On July 2, 1644, Gascoigna died at the Battle of Marston Moor in Yorkshire.

Gödel. Dr. Kurt Gödel (1906-1978) was an Austro-Hungarian-born mathematician and logician. He had an immense effect upon scientific thinking in the 20th century. He studied theoretical physics, mathematics, and philosophy. He was good friends with Albert Einstein. Later in life, he suffered periods of mental instability and illness. He had an obsessive fear of being poisoned. He would eat only food that his wife cooked, and only after she tasted it first. In late 1977, she was hospitalized for 6 months and was unable to prepare her husband's food. In her absence, he refused to eat anything, and eventually starved to death on January 14, 1978. He weighed on 65 pounds when he died. [Source: Hamilton, "Kurt Gödel – The Brilliant, Paranoid Mathematician who Refused to Eat," *Vintage News*, April 24, 2018]

Goldmark. Peter Carl Goldmark (1906-1977) was a Hungarian-American scientist and engineer who developed the long-playing 33 1/3 rpm phonograph disc. He also pioneered color television while at CBS. Cameras using his color wheel system was used on the lunar surface TV cameras during the Apollo moon landings. Goldmark also invented the Electronic Video Recorder in 1967. On November 22, 1977, President Jimmy Carter presented Goldmark with the National Medal of Sciences for all of his contributions. 15 days later, on December 7, 1977, Goldmark died in an automobile accident in Harrison, Westchester County, New York. His car struck another vehicle on the Hutchinson River Parkway. He was driving alone on his way to an airport to catch a flight to Los Angeles for a business meeting. [source: Stahlberg, "Inventor's death called blow to major school advances," *Eugene Register-Guard*, Dec 8, 1977]

Grubisic. Dr. Angelo Grubisic (1981-2019) was a British space scientist and astronautical engineer who designed safer wingsuits. He had previously worked for NASA on spacecraft propulsion. In August 2019, he died in Saudi Arabia during a BASE jumping incident. [source: "Wingsuit scientist dies in Saudi Arabia base jump," *BBC News*, Aug 22, 2019]

Hawkins. Angela Nicole Chadwick-Hawkins was a wildlife biologist at Fort Jackson, South Carolina. She was killed during a controlled fire on the military post on May 22, 2019. A blaze was set intentionally near her to clear brush and improve wildlife habitat. Her charred body was found near an ATV she had ridden that day. [source: Fretwell, :Big Mystery. Unusual Fort Jackson fire death has biologist's family asking 'Why?'" *thestate.com*, June 11, 2019]

Johnston. David A. Johnston (1949-1980) was an American volcanologist and a principal scientist with the United States Geological Survey (USGS). He was killed on May 18, 1980 by the eruption of Mount St. Helens in Washington State. He was 6 miles away from the eruption and was the first to report the eruption. He was swept away by a lateral blast of hot ash. His body was never found. [source:

Oddities of Science

Hunter, "The Cataclysm: 'Vancouver! Vancouver! This Is It!'" *Scientific American*, August 9, 2012]

Kildall. Dr. Gary Kildall (1942-1994) was an American computer scientist who created the CP/M operating System and founded Digital Research. In July 1984, he fell at a Monterey, California, biker bar and hit his head, suffering a concussion. Various sources claimed he fell from a chair, fell down steps, or was assaulted in a biker bar brawl. He checked into and out of a hospital twice, but died 3 days later on July 11, 1994 at a hospital in Monterey. [source: Schestowitz, "The Death of Gary Kildall Remains a Mystery to This Date," *techrights.org,* Jan 5, 2020]

Lapenta. William Lapenta (1961-2019) was the director of the National Oceanic & Atmospheric Administration's (NOAA) National Centers for Environmental Prediction. In 2019, he drowned in rough surf off North Carolina's Outer Banks. [source: Price, "Prominent scientist with National Weather Service dies in waters off Outer Banks," *Charlotte Observer*, Oct 2, 2019]

Midgley. Thomas Midgley, Jr (1889-1944) was an American chemist and mechanical engineer. He played a major part in developing leaded gasoline (with Charles Kettering) and CFCs (Freon). He was awarded over 100 patents in his lifetime. In 1940, he contracted polio, which left him severely disabled as he was paralyzed below the waist. He later created a system of strings and pulleys to lift himself out of bed. On November 2, 1944, he became entangled in the ropes and was strangled by them. Midgley's legacy may be that he was the most environmentally disastrous person of all time with his development of leaded gasoline and Freon. He contributed to the poisoning of 3 generations of children, increased the risk of skin cancer and other skin problems related o exposure of ultraviolet rays, and contributed greatly to global warming. [source: Thomas Midgley Jr." The Man Who Harmed the World the Most," *interestingengineering.com*, Aug 6, 2018]

Moseley. Henry Moseley (1887-1915) was an English physicist. In 1912, he invented the first atomic battery using radium. In 1913, he discovered that every chemical's identity is determined by its number of protons. He discovered the true basis of the periodic table. When World War I began in 1914, he enlisted as a volunteer as a telecommunications (signals) officer for the Royal Engineers of the British Army. On August 10, 1915 2nd Lt. Henry Moseley was shot in the head and killed by a sniper in Gallipoli, Turkey while in the act of telephoning a military order. [source: Patton, "Henry G.J. Moseley – Battle Death at Gallipoli of Promising Young Physicist," *kumc.edu*, April 8, 2019]

Parsons. John "Jack" Parsons (1914-1952) was an American rocket engineer, rocket propulsion researcher, and chemist. He was a co-founder of the Jet Propulsion Laboratory (JPL) and the Aerojet Engineering Corporation. He was also a leader of a black-magic sex cult, of which Scientology founder L. Ron Hubbard was once a member. In 1952, he travelled to Mexico to establish an explosives factory for the Mexican government. On June 17, 1952, he received a rush order of explosives for a film set. He began to work on it in his home laboratory when an explosion blew up the lower part of his house, during which Parsons sustained mortal wounds. The explosion blew off his right forearm. His legs and left arm were broken. A hole was torn in the right side of his face. Despite these critical injuries, he was found conscious, but died in an ambulance on the way to a hospital.  His mother was so distraught when she learned of her son's death that she immediately took a fatal overdose of barbiturates. [source: Oleksinski, "This sex-crazed cultist was the father of modern rocketry," *New York Post*, June 19, 2018]

Rawlings. Steven Rawlings (1961-2012) was an astrophysicist at the University of Oxford. He died of cardiac arrest at his home after being put in a headlock by a colleague, Dr. Devinder Sivia. The two got into a fight and Sivia was trying to restrain Rawlings. Sivia was arrested on suspicion of murder but later released without charge. [source: *The Guardian*, Nov 28, 2012]

Oddities of Science

Reichelt. Franz Reichelt (1878-1912) was an inventor and parachuting pioneer. On February 4, 1912, he climbed to the top of the Eiffel Tower in a crazy wingsuit of his own design, a wearable parachute. Friends who accompanied him tried convincing him to first throw a dummy to see if the suit actually works. But Reichelt refused. Once at top of the first platform of the Eiffel Tower, he adjusted the suit, faced the Seine, tested the wind direction, hesitated for about 40 seconds, then jumped. He was observed by 30 journalists, two cinematographers (one on top and one on the ground), and a large crowd.  His parachute folded around Reichelt almost immediately as he fell 187 feet to his death, hitting the frozen ground head first that left a crater almost 6 inches deep. An autopsy later determined that he died of a heart attack during the fall. Reichelt's death was the first to result from a parachuting accident since Charles Leroux died giving a demonstration of his parachute in 1889. Just two days prior to Reichelt's jump, an American parachute enthusiast made a successful jump from the Statue of Liberty with a wearable parachute. In 1930, the first successful wingsuit was used by Rex Finney in Los Angeles. It wasn't until the mid-1990s that the modern wingsuit was developed.  [source: Franz Reichelt: Dressmaker Who Jumped off the Eifel Tower Experimenting with Self-Designed Parachute Suit," *statworld.com*, Aug 26, 2019]

Scheele. Carl Scheele (1742-1786) was a Swedish-German pharmacist and chemist. He discovered 8 elements – chlorine, fluorine, manganese, barium, molybdenum, tungsten, nitrogen, and oxygen. He also discovered ammonia, glycerin, tannic acid, citric acid, hydrogen cyanide, and other compounds. He was also the first person to isolate lactic acid from sour milk. He had an odd habit of smelling and tasting a little of everything he worked with, which may have killed him. On May 21, 1786, at the age of 43, he was found dead in his lab workbench, surrounded by an array of toxic chemicals, such as hydrocyanic acid, hydrofluoric acid, arsenic, lead, and mercury. He had gotten married two days earlier. Doctors said that he died of mercury poisoning and suffered from kidney disease. [source: Patowary, "Carl Wilhelm Scheele: The Unlucky Chemist," *Amusing Planet*, Oct 22, 2019]

Slotin. Dr. Louis Slotin (1910-1946) was a Canadian physicist and chemist. He worked on the Manhattan Project. He performed experiments with uranium and plutonium cores to determine their critical mass values. He helped assemble the first atomic bomb. On May 21, 1946, near Los Alamos, New Mexico, he was experimenting with two half-spheres of beryllium that was acting as a neutron reflector around a plutonium core. He was using a screwdriver to separate the two halves of the beryllium sphere. The screwdriver slipped and the upper beryllium hemisphere fell onto the plutonium core, causing a prompt critical reaction and a big burst of radiation. Slotin had been exposed to a lethal dose of radiation (2,100 rem when 500 rem is usually fatal for humans) and died 9 days later. He was buried in a sealed Army casket. [source: Wellerstein, "The Demon Core and the Strange Death of Louis Slotin," *The New Yorker*, May 21, 2016]

Tesla. Nikola Tesla (1856-1943) was an inventor, electrical engineer, and mechanical engineer. He died on January 7, 1943 in his room at the New Yorker Hotel, poor and reclusive. The cause of death was coronary thrombosis. He few years earlier, he had been hit by a moving taxicab, breaking 3 ribs and severely wrenching his back. He never recovered from his injuries. After his death, the FBI ordered that all of Tesla's belongings and papers be seized in the interests of national security. Though most of it was later returned to his family, some of it still remains classified. His ashes are displayed in a gold-plated sphere on a marble pedestal in the Nikola Tesla Museum in Belgrade. [source: "Nikola Tesla," *biography.com*, Sep 4, 2019]

Ward. Mary Ward (1827-1869) was an Irish naturalist and astronomer. On August 31, 1969, she was killed when she fell under the wheels of an experimental steam car built by her cousins. The vehicle jolted during a turn, and Ward was bucked off. When she hit the ground, one of the huge iron wheels crushed her. The fatality was a broken neck. She was the first person known to have been killed by a motor vehicle. [source: Kean, "The First Car-Crash Victim was a Female Scientist," *The Atlantic*, Aug 29, 2019]

Oddities of Science

Wegener. Alfred Wegener (1880-1930) was a German geophysicist and meteorologist. He was the originator of the theory of continental drift. In 1906, he and his brother set a new record for a continuous balloon flight, remaining aloft for over 52 hours, while carrying out meteorological investigations. In September 1930, he set out with 13 Greenlanders and his meteorologist to supply one of three scientific camps. In November 1930, Wegener died while trying to bring food to a group of researchers at another camp called West camp. His body was found 6 months later.

Zhankov. Alexander Zhankov was a Russian professor at Oxford University's Hydrology and Ecology Center. What he thought was a beaker of ethanol was a beaker of methanol instead. Methanol is 5 times more toxic and the professor died a slow, lingering death on a life support system. [source: Payne, "Drink error killed professor," *The Telegraph*, May 18, 2001]

Suicides of Scientists

Addison. Thomas Addison (1793-1860) was a scientist and medical doctor. He discovered what is known as Addison's disease, a degenerative disease of the adrenal glands. He suffered from many episodes of marked depression and mental illness. On June 29, 1860, he committed suicide in Brighton by eluding two attendants and jumping out of the top story window where he resided. He fell 9 feet, falling on his head. [source: Cirillo, "The Suicide of Thomas Addison," *Journal of the History of Medicine and Allied Sciences*, April 1, 1985]

d'Archiac. Adolphe d'Archiac (1802-1868) was a French geologist and paleontologist. He described many of the geological formations of France, Belgium, and England. He studied the distribution of fossils. While suffering from severe depression, he committed suicide by throwing himself into the River Seine on Christmas Eve in 1868. [source: "Adolphe d'Archiac: French paleontologist." *peoplepill.com*]

Barth. Dr. Hermann von Barth (1845-1876) studied natural sciences and was a famous German mountaineer, climbing over 100 peaks, and the first to climb over a dozen peaks. On December 7, 1876, deranged with fever, he committed suicide while on a research expedition in Angola, Africa. [source: *The Popular Science Monthly*, 1877, p. 128]

Berger. Hans Berger (1873-1941) was a German medical doctor and psychiatrist. He invented the electroencephalography (EEG) to measure brain waves. There is some evidence that he was a member of the Nazi SS and served on the "Court for Genetic Health" that imposed sterilizations. He suffered from a long period of clinical depression, chronic heart problems, and severe skin infections. On June 1, 1941, he hanged himself at his clinic. [source: Ibanez, "Hans Berger: Lights and Shadows of the Inventor of Electroencephalography," *neuroelectrics.com*, Dec 18, 2014]

Berliner. Arnold Berliner (1862-1942) was a German physicist and a Jew. He was the editor and founder of a prestigious scientific magazine. In 1935, the Nazi government dismissed Berliner as the editor because of their racial policies on "non-Aryans." On March 22, 1942, he committed suicide the day before he was to be deported to an extermination camp. He took poison that night, probably hydrocyanic acid and succumbed to nervous vomiting. [source: "Arnold Berliner," *mahlerfoundation.org*, Nov 21, 2019]

Bettelheim. Dr. Bruno Bettelheim (1903-1990) was an Austrian-born American psychologist. In 1938, he was arrested and sent, first to Dachau concentration camp, and then to Buchenwald. His family succeeded in getting him released, and he left for New York. He was a world expert in autism and child psychiatry. He was considered cruel and believed in corporal punishment in the therapeutic setting, which would not be acceptable today. He suffered from depression for years and planned his suicide for over a year. On March 13, 1990, he committed suicide by placing a plastic bag over his head and died of asphyxiation. [source: Green, "An Imperfect Psychoanalyst Commits Suicide,: Haaretz, March 13, 2016]

Boltzmann. Ludwig Boltzmann (1844-1906) was an Austrian physicist. He is known for the Boltzmann constant, a physical constant that relates the average relative kinetic energy of particles in a gas with the temperature of the gas. In 1906, he was suffering from a deteriorating mental condition. On September 5, 1906, he hanged himself while on vacation with his wife and daughter in Austria.

Bridgman. In 1946, American physicist Dr. Percy Williams Bridgman (1882-1961) won the Nobel Prize in Physics for his work on the

Oddities of Science

physics of high pressures. In his later years, he was suffering from metastatic cancer. On August 20, 1961, he committed suicide by gunshot. His suicide note read in part, "It isn't decent for society to make a man do this thing himself. Probably this is the last day I will be able to do it myself."

Büyükkasap. Erdoğan Büyükkasap (1962-2010) was a Turkish scientist and president of Erzincan University in Turkey. On March 18, 2010 he hanged himself at his home.

Campbell. William Wallace Campbell (1862-1938) was an American astronomer and director of the Lick Observatory in California. He was a pioneer in astronomical spectroscopy. He was President of the University of California from 1923 to 1930. Unwilling to be a burden on his family, he committed suicide by leaping to his death from his 4th story window in San Francisco. [source: "Dr Campbell Dies in Leap on Coast," *New York Times,* June 15, 1938]

Carothers. Dr. Wallace Carothers (1896-1937) was an American chemist. He was the leader of organic chemistry at DuPont, and is credited with the invention of nylon in 1934. He was the first organic chemist elected to the National Academy of Sciences. He suffered from depression. He committed suicide in a hotel room in Philadelphia by drinking potassium cyanide dissolved in lemon juice. [source: "A Science Odyssey: People and Discoveries: Wallace Carothers, *PBS.org*]

Clark. Dr. Anthony John Clark (1951-2004) was an English molecular biologist. He was a founder of applying molecular technology to farm animals. Through his genetic modification, he created a sheep that produced large quantities of human protein, used in the treatment of cystic fibrosis. His work led to the technique of cloning Dolly the sheep in 1996. He suffered from depression, and on August 12, 2004, he was found hanged in his holiday home. [source: Wilmut, "John Clark," *The Guardian*, Aug 25, 2004]

Crookshank. Dr. Francis Crookshank (1873-1933) was a British epidemiologist and a Fellow of the Royal College of Physicians. He wrote two scientific racist publications on eugenics. On October 27, 1933, he committed suicide at his house in London.

Denton. Denice Denton (1959-2006) was an electrical engineer and the chancellor of the University of California, Santa Cruz. She was the first woman in the nation to lead an engineering college at a major university, and was the youngest among UC's 10 chancellors. On June 24, 2006, one day following her discharge from a clinic that was treating her for depression, she leapt 33 stories to her death from a high-rise apartment where she was living in San Francisco. [source: "Denice Dee Denton Obituary," San Jose *Mercury News*, June 24, 2006]

Derby. Orville Derby (1851-1915) was an American geologist who worked in Brazil. He organized the first geological commission in Brazil and found the first botanical gardens in Sao Paulo. He published 173 papers on geology and published one of the first geological maps on Brazil in 1915. On November 15, 1915, he committed suicide in a hotel room in Rio de Janeiro. [source: "O. A. Derby A Suicide," *New York Times*, Nov 29, 1915]

Drude. Paul Drude (1863-1906) was a German physicist specializing in optics. He was the editor of the most respected physics journal at that time. In 1894, he was responsible for introducing the symbol "c" for the speed of light in a perfect vacuum. In July 1906, he was elected a member of the Prussian Academy of Sciences. A few days later, on July 5, 1906, he committed suicide in his laboratory by poisoning himself with prussic acid. [source: Dieter, "Paul Drude," *Annalen der Physik*, July 2006]

Eady. Eric Eady (1915-1966) was a British meteorologist. He was an expert on atmospheric instability and the development of weather systems. In his later years, he became depressed by his career. On September 5, 1915, he purposely took an overdose of sleeping pills while a patient at the Royal Surrey County Hospital.

Ehrenfest. Paul Ehrenfest (1880-1933) was an Austrian and Dutch theoretical physicist. His major contributions were in the field of statistical mechanics and quantum mechanics. He was a good friends of Albert Einstein and Niels Bohr. He suffered from severe depression. On September 25, 1933, he fatally shot his younger son Wassik, who had Down syndrome, then shot himself. His son was a patient at the Waterink Institute for Afflicted Children in Amsterdam.   [source:

Oddities of Science

Halpern, "The Tragic Fate of Physicist Paul Ehrenfest," *medium.com*, Feb 10, 2015]

Fischer, E. Dr. Emil Fischer (1852-1919) was a German organic chemist who won the 1902 Nobel Prize in Chemistry. He was instrumental in the discovery of caffeine and barbiturates, and invented the first barbiturate sedative, barbital in 1904. He was suffering from terminal intestinal cancer. On July 15, 1919, he committed suicide in Berlin. [source: "Emil Fischer," *Science History*, Dec 5, 2017]

Fischer, H. Hans Fischer (1881-1945) was a German organic chemist who won the 1930 Nobel Prize in Chemistry. He was mostly concerned with the investigation of the pigments of blood, bile, and chlorophyll in leaves. He suffered from tuberculosis. On March 31, 1945, he committed suicide in Munich in despair over the destruction of his institute during the last day of World War II. [source: "Hans Fischer Biography," *thefamouspeople.com*, Oct 25, 2017]

FitzRoy. Robert FitzRoy (1805-1865) was a meteorologist and a Vice Admiral in the British Navy. He was the captain of the *HMS Beagle* that Charles Darwin accompanied as a companion. He founded the British Meteorological Department and created a weather prediction office. He coined the word 'forecasting' for coming weather. He was the first to telegraph weather forecasts. He suffered from depression. On April 30, 1865, he committed suicide by cutting his throat with a razor. He died penniless after spending all of his money on public meteorological projects. [source: Moore, "The Tragic Life of Charles Darwin's Captain," *History Today*, May 28, 2015]

Goodall. David Goodall (1914-2018) was an English born Australian botanist and ecologist. His major work was on the development of statistical methods in plant communities. He authored over 100 publications. He was known as Australia's oldest working scientist, still editing ecology papers at age 103. He was an advocated of voluntary euthanasia and died by physician-assisted suicide using a lethal injection in Switzerland at the age of 104 on May 10, 2018. He was not terminally ill. [source: Steinbuch, "Scientist who regretted living to 104 dies in assisted suicide," *New York Post*, May 10, 2018]

Hardin. Garrett Hardin (1915-2003) was an American zoologist, microbiologist, and ecologist. His major focus was on the issue of human overpopulation. He suffered from a heart disorder and polio. His wife suffered from Lou Gehrig's disease. On September 14, 2003, he and his wife committed suicide shortly after their 62nd wedding anniversary. [source: Lavietes, "Garrett Hardin, 88, Ecologist Who Warned About Excesses," New *York Times,* Oct 28, 2003]

Hashida. Kunihiko Hashida (1882-1945) was a Japanese physician and physiologist. He served as Minister of Education from 1940 to 1943 in Japan. After World War II, he was accused of war crimes, which he denied. On September 14, 1945, he committed suicide by taking potassium cyanide so as to avoid facing trial by the Allies.

Hughes. Nicholas Hughes (1962-2009) was an English American zoologist and biologist known as an expert in stream salmonid ecology. He was the son of American poet Sylvia Plath and English poet Ted Hughes. He suffered from depression. On March 16, 2009, he hung himself in his home in Fairbanks, Alaska. His mother committed suicide by gas when he was a child. [source: O'Connor, "Nicholas Hughes, 47, Sylvia Plath's Son, Dies," New York Times, March 24, 2009]

Immerwahr. Clara Immerwahr Haber (1870-1915) was a German chemist. She was the first woman to be awarded a doctorate in chemistry in Germany. She was married to Nobel Prize-winning chemist Fritz Haber. On May 2, 1915, she shot herself in the chest in their garden using her husband's pistol. A week before her death, her husband had organized the first chlorine-gas (200 tons) attack of World War I at Ypres, Belgium. [source: "Casualty of War," *Science History Institute*, July 5, 2012]

Ivins. Bruce Edwards Ivins (1946-2008) was an American microbiologist. He was a senior bio-defense researcher at Fort Detrick, Maryland. He was a suspected perpetrator of the 2001 anthrax attacks that killed 5 people. On July 29, 2008, he died by suicide after taking an overdose of Tylenol with codeine two days earlier. He had just learned that criminal charges were likely to be filed against him by

the FBI in connection with the 2001 anthrax attacks. However, no formal charges were ever filed and there was no direct evidence of his involvement in the anthrax attacks. The Anthrax investigation cost the government $100 million. [source: Engelberg, "New Evidence Adds Doubt to FBI's Case Against Anthrax Suspect," *ProPublica*, Oct 10, 2011]

Kelly. Dr. David Christopher Kelly (1944-2003) was a Welsh scientist and authority on biological warfare. He had gone for a walk and ingested 29 tablets of co-proxamol, an analgesic drug, then cut his left wrist with a pruning knife. He died underneath a tree at Harrowdown Hill in Oxfordshire, England. Some sources say that he was murdered. [source: "Dr. David Kelly: 10 years on, death of scientist remains unsolved for some," *The Guardian*, July 16, 2013]

Kohlberg. Lawrence Kohlberg (1927-1987) was an American psychologist and professor of education at Harvard University. His specialty was on the stages of moral development and moral education. He suffered from depression and a parasitic infection he caught while doing research in Belize. On January 17, 1987 he committed suicide by jumping into the icy Boston Harbor and drowned. His body was found a few weeks later. [source: "Lawrence Kohlberg Is Dead," *New York Times*, April 8, 1987]

Lange. Andrew Lang (1957-2010) was an astrophysicist whose interests focused on the Cosmic Microwave Background (CMB). He developed a new generation of radio receivers for this study. He was chair of the Division of Physics, Mathematics and Astronomy at the California Institute of Technology. He was married to Frances Arnold, winner of the Nobel Prize in Chemistry in 2018. On January 21, 2010, he checked into a hotel in Pasadena and committed suicide by asphyxiation.

Leblanc. Nicolas Leblanc (1742-1806) was a French chemist and surgeon. He discovered how to manufacture soda ash (sodium carbonate) from common salt. He built a plant that produced 320 tons of soda ash per year. Two years later, the French revolutionary government confiscated the plant. In 1802, Napoleon returned the plant back to Leblanc. However, Leblanc could not afford to run it. On January 16, 1808 he committed suicide by shooting him-

self in the head. [source: Sutton, "A revolutionary casualty," *Chemistry World*, October 30, 2006]

Legasov. Valery Legasov (1936-1988) was a Soviet inorganic chemist and a member of the Academy of Sciences of the USSR. He was a professor at the Moscow Institute of Physics and Technology. He was chief of the commission investigating the Chernobyl disaster which occurred in April 1986. On April 27, 1988, one day before he was due to announce his results of the investigation of the Chernobyl Nuclear Power Plant accident, he died by suicide by hanging himself in his apartment. [source: Bilyeau,"The Man Who Investigated Chernobyl – The Sad Fate of Valery Legasov," *Vintage News*, June 6, 2019[

McDonald. James Edward McDonald (1920-1971) was a meteorologist and senior physicist at the Institute for Atmospheric Physics in Tucson, Arizona. He was an expert on atmospheric phenomena and cloud formation. He is best known for his research regarding UFOs. He interviewed over 500 UFO witnesses and uncovered many important government UFO documents. He testified before Congress during the UFO hearings of 1968. In April 1971, he attempted suicide by shooting himself in the head. He survived, but was blinded. On June 13, 1971, he shot himself near a creek in Tucson. A .38 caliber revolver was found close to him, as well as a suicide note. [source: "James M'Donald, A Cloud Physicist," *New York Times*, June 16, 1971]

McKinstry. Chris McKinstry (1967-2006) was a Canadian researcher in artificial intelligence. He was an operator of the Very Large Telescope in Chile. He suffered from bipolar disorder. He wrote two suicide notes on the Internet and on January 23, 2006, he was found dead in his Santiago apartment with a plastic bag over his head and a hose that was connected to the gas pipe.

Métraux. Alfred **Métraux** (1902-1963) was a Swiss and Argentine anthropologist and ethnologist. He published landmark studies of South American Indians and the history of the Incas. He solved the mysteries of the stone statues of Easter Island. He taught the world about Voodoo in Haiti. On April 12, 1963, he committed suicide by taking an overdose of barbiturates. At the time, he was Professor of South American Anthropology in Paris.

Meyer. Dr. Viktor Meyer (1848-1897) was a German chemist. He invented an apparatus for determining vapor densities. He had a major nervous breakdown during the last years of his life. On the night of

Oddities of Science

August 7-8, 1897, he committed suicide in Heidelberg, Germany, by taking cyanide. [source: "Obituary: Victor Meyer," *The Chemical News and Journal of Physical Science*, Volume 76, 1897, p.106]

Mukhodpadhyay. Subhash Mukhopadhyay (1931-1981) was an Indian Bengali scientist and physician. In 1978, he created the world's second and India's first child using in-vitro fertilization ("test tube baby"). Afterwards he was harassed by the West Bengal and Indian government. They would not allow him to share his achievements with the international scientific community and they refused to allow him to attend international conferences. Dejected and ridiculed, he committed suicide hanging himself in Calcutta at his home on June 19, 1991, leaving a suicide note. [source: "Indian Medical Association under fire over Calcutta physiologist's death." *India Today*, Oct 15, 2014]

Nelson. Edward Nelson (1883-1923) was a British marine biologist and polar explorer. In 1910, he joined the British Antarctic Expedition, led by Robert Scott and served as a biologist. On January 17, 1923, he committed suicide in his laboratory in Aberdeen, Scotland, by injecting himself with poison. His daughter, Barbara, then age 93, died during a trip to Antarctica in 2009.

Von Neumann, Klara. Klara Dan von Neumann (1911-1963) was a Hungarian-American computer scientist. She was one of the first computer programmers. In 1938, she married John von Neumann. In 1945, she designed new controls for the ENIAC computer and was one of its primary programmers. She was a key figure in the experiment that launched modern weather prediction. On November 10, 1973, she committed suicide by walking into the surf in San Diego and drowned. [source: Witman, "Meet the Computer Scientist You Should Thank For Your Smartphone's Weather App," *Smithsonian Magazine*, June 16, 2017]

Nopcsa von Felso-Szilvas. Franz Nopcsa von Felso-Szilvas (1877-1933) was an Austro-Hungarian geologist and paleontologist. He was one of the founders of paleobiology and studied small dinosaurs. He developed the theory that birds evolved from dinosaurs, which later proved to be true. He suffered from depression and a nervous breakdown. On April 25, 1933, he fatally shot his partner, Bajazid Doda, after having laced his tea with sleeping powder. He then shot himself. [source: Osterloff, "Franz Nopcsa: the dashing baron who discovered dwarf dinosaurs," *Natural History Museum*, Feb 27, 2019]

Northrop. John Howard Northrop (1891-1987) was an American biochemist. In 1938, he isolated and crystallized the first bacteriophage (a virus that attacks bacteria). He shared the 1946 Nobel Prize in Chemistry for his work in the study of enzymes, proteins, and viruses. On May 27, 1986, at the age of 95, he committed suicide in Wickenburg, Arizona.

Von Pirquet. Clemens von Pirquet (1874-1929) was an Austrian scientist and pediatrician. He coined the word allergy to describe hypersensitive reaction. He developed a test for tuberculosis. He was nominated 5 times for a Nobel Prize, but never won. On February 28, 1929, he and his wife committed suicide with potassium cyanide.

Price. George Robert Price (1922-1975) was an American chemist and evolutionary biologist. He was a member of the Manhattan Project as a chemist looking into the characteristics of plutonium 235. He came up with the Price equation, which describes how a trait or gene changes in frequency over time. In 1973, he invited homeless people around Soho Square to live in his home. After his lease ran out, he became homeless himself. On January 6, 1975, he committed suicide by cutting his carotid artery in his neck with a pair of nail scissors. [source: Matthews, "The scientist who tried to be as selfless as possible, until it killed him," *Vox*, Jan 6, 2019]

Sasai. Yoshiki Sasai (1962-2014), a Japanese stem-cell scientist, hanged himself in the stairwell of the RIKEN Center for Developmental Biology (CDB) in Kobe, Japan. He had got caught up in a stem-cell scandal in which two of his research papers that he co-authored were retracted. The lead author, Haruko Obokata, was found guilty of scientific misconduct. Sasai was found innocent of any direct involvement with the problematic data. [source: Cyranoski, "Stem-cell pioneer blamed media 'bashing' in suicide note," *Nature*, Aug 15, 2014]

Tovey. In 2008, Francis Pete Tovey, age 81, of Australia, killed himself using a robot he built himself. The robot consisted of a .22 caliber semi-automatic pistol, a jigsaw power tool, and 4 bullets that could fire multiple shots. He rolled out his robot onto his driveway at 7 am and set the thing off. It worked, killing him with three shots to the head. [source: "Australian kills himself with suicide machine," *The Telegraph*, Mar 20, 2008]

Oddities of Science

Turing. Alan Turing (1912-1954) was a computer scientist, mathematician, and cryptanalyst. He played a crucial role in cracking intercepted messages during World War II. In 1952, he was prosecuted in Britain for homosexual acts. He accepted chemical castration treatment instead of prison. On June 7, 1954, he probably committed suicide by cyanide poisoning eating a cyanide-laced apple. Other sources report that he died of accidental inhalation of cyanide fumes from an apparatus he had in his home used to electroplate gold onto spoons. [source: Copwell, "Overlooked No More: Alan Turing, Condemned Code Breaker and Computer Visionary," *New York Times,* June 5, 2019]

Wells. Horace Wells (1815-1848) was an American dentist who pioneered the use of anesthesia in dentistry in 1844, specifically, the use of nitrous oxide (or laughing gas). He later became addicted to chloroform. On January 21, 1848, on his 33rd birthday, he rushed into the street in Manhattan and threw sulfuric acid on two prostitutes. He was arrested and committed to Tombs Prison in New York. Three days later, he asked the guards to escort him to his house to pick up his shaving kit. Upon returning to his cell, he committed suicide by slitting an artery with a razor after inhaling some chloroform. [source: "Horace Wells and Anaesthesia," *Nature*, Jan 24, 1948]

Scientists Who Disappeared

Alexandrov. Vladimir Alexandrov (1938-?) was a Soviet physicist who created a mathematical model for the nuclear winter theory. He was a pioneer in global climate modelling who had access to a Cray-1 supercomputer. He disappeared from Madrid, Spain, on the evening of March 31, 1985, after attending the Second International Conference of Nuclear Free Zones in Cordoba, Spain. On theory was that he defected. Another theory was that he was kidnapped by the KGB. He left behind a wife and daughter back in Moscow. [source: Ansede & Sahuquillo, "The Soviet scientist who disappeared in 1980s Madrid," *El Pais,* Jan 14, 2019]

Bista. Dor Bahadur Bista (1924-1995?) was a Nepalese anthropologist and considered the father of Nepalese anthropology. He was also a crusader against the caste system and made enemies with the elite.

The Nepal government tagged him as an 'anti-national.' In June 1995, he was last seen boarding a bus in western Nepal. There has been no sign of him since. [source: Dixit, "Looking for Dor Bahadur Bista," *Nepali Times*, April 17, 2015]

Cogar. George Cogar (1932-1983?) was a computer scientist and the head of the UNIVAC 1004 electronic design team. He invented the Data Recorder magnetic tape encoder, which made him a millionaire. He was last seen on September 2, 1983, when a private plane that he flying went down somewhere in British Columbia, Canada. The plane, Cogar, and 5 passengers, has never been found. After a week, searchers had been in the air for almost 800 hours and covered 32,000 square miles. It was the largest coordinated search in Canada's history, costing about $1 million.

Fawcett. Percy Fawcett (1867-?) was a British archeologist and explorer of South America. In 1925, he, and his son, another companion, two laborers, and a ton of supplies disappeared during an expedition looking for the ancient lost city of Z in the jungles of Brazil. The last communication from is expedition was on May 29, 1925, when he wrote a letter to his wife delivered by a native runner. Fawcett sent dispatches of his exploration, which began in February 1925. His journey was published in most North American newspapers and he had 40 million readers following along. His party may have been  killed by local Indians. Another theory is that his party got lost and starved to death. His lost city was found. Years later, archeologists discovered huge cities of stone fortified by walls hidden within the jungle, not too far from where Fawcett party disappeared. Fawcett's adventures inspired the Indian Jones novel. [source: Berman, "The True Story Behind the Lost City of Z," *Time*, April 14, 2017]

Gray. Dr. James Gray (1944-2007?) was an American computer scientist. In 1998, he received the Turing Award for his contributions to database and transaction processing research. In 1995, he joined Microsoft and was a Technical Fellow. He was an experienced sailor and owned a 40-foot sailboat. On January 28, 2007, he failed to return from a short solo trip to the Farallon Islands, 30 miles off of the San Francisco coast. The weather was clear and there was no distress call. For several days, the Coast Guard searched 132,000 square miles with planes, helicopters, and boats, but found nothing. NASA volunteered to steer a high-altitude aircraft over the search area. Satellite imagery using Google Earth was also examined. His wife hired an underwater search team, and they found nothing. He and his sailboat disappeared.

Grinberg. Dr. Jacobo Grinberg (1946-1994?) was a Mexican neurophysiologist and psychologist. He was the author of over 50 scientific books. On December 12, 1994, his family prepared a party for him to celebrate his 48th birthday, but he did not show up. He had planned on a trip to Nepal. There has been no trace of him. [source: Quinones, "Looking for Doctor Grinberg," *New Age Journal*, July/August 1997]

Krug. Dr. Heinz Krug (1913-1962?) was a German scientist and was a Nazi rocket expert during World War II at Peenemunde where he worked with Wernher von Braun. Krug was offered a job in American, working for von Braun, but Krug turned it down. On September 11, 1962, we went to his office in Munich and never came home. He was last seen in the company of an Arab named Saleh. A few days later, Krug's car was found abandoned. He was hired by the Egyptian government to develop advanced weapons for Egypt. It was first reported that Krug was kidnapped by the Egyptians to prevent him from doing business with Israel. Based on interviews with former Mossad officers, Krug was murdered as part of an Israeli espionage plot to intimidated the German scientists working for Egypt. [source: Meiman, "The Strange Case of a Nazi Who Became an Israeli Hitman," *Haaretz*, March 27, 2016]

Majorana. Ettore Majorana (1906-1959?) was an Italian theoretical physicist. He was the first to interpret correctly the experiment that a neutron had a neutral charge and the mass about the same as a proton. Majorana did not bother to publish his findings. It was left up to Sir James Chadwick (1891-1974) to publish a paper on the neutron in 1932, which won him a Nobel Prize in Physics in 1935. He predicted neutrinos and a class of fermion antiparticles. On March 25, 1938, he disappeared during a boat trip from Palermo to Naples. His body was never found and his fate is still uncertain. Before he disappeared, he withdrew all his money from his bank account. On the day of his disappearance he sent a note to a colleague who was the Director of the Naples Physics Institute. He wrote that his sudden disappearance would cause trouble with the director and the students. In 2015, the Rome Attorney's Office released a statement declaring that Majorana was alive between 1955 and 1959, and living in Venezuela. These last findings, base on witnesses and new evidence, made the Rome Attorney's Office to declare the case officially closed. [source: "The Mysterious Disappearance of Ettore Majorana, Finally Solved" *mysteriousfacts.com*, March 8, 2020]

Mutch. Thomas Mutch (1931-1980?) was an American geologist and planetary scientist. He was a geology professor at Brown University in Providence, Rhode Island. He published two books about the geology of the Moon and of Mars. He was an Associated Administrator for Space Science at NASA. He was head of the Mars Viking 1 photography team in the 1970s.  In October 1985, he climbed Mount Nun (23,409 feet) in the Kashmir Himalayas. He disappeared during his descent from the mountain (most mountaineering accidents happen on the way down). He was last seen on a ledge just after losing a crampon. He could not get down without the crampon, so one of his two companions went to find a spare. When he returned, Mutch was gone. His body was never found. NASA renamed the Viking 1 Lander the "Thomas A. Mutch Memorial Station." [source: Imbler, "Professor leaves legacy of exploration," *Brown Daily Herald*, May 24, 2013]

Peng Jiamu. Peng Jiamu (1925-1980?) was a Chinese biochemist. He worked at the Shanghai Institute of Biochemistry and Cell Biology. He catalogued species of flora and fauna in the Lop Nor desert (nicknamed Asia's Devil's Triangle) in China. In 1980, he led a scientific expedition in the Lop Nur desert. On June 17, 1980, he left a note in his camp saying he had gone to find water. He was never seen again. Chinese leaders sent planes, helicopters, police, dogs, and hundreds of soldiers to scour the landscape, but Peng had completely vanished. [source: Lloyd, "Mysterious Disappearance of Explorer Peng Jiamu In Lop Nur – the Wandering Lake," *ancientpages.com*, Dec 20, 2018]

Philip. Sneha Anne Philip (1969-2001?) was an American physician working in New York City at the Jacobi Medical Center. She was last seen on September 10, 2001 in a department store in Lower Manhattan. Due to the proximity of the World Trade Center and her medical training, her family believes she perished trying to help victims of the following day's terrorist attacks on the World Trade Centers. There were several investigations into her disappearance. One investigation showed that she had a double life, a history of mental problems, and possible affairs, along with job difficulties and alcohol and drug abuse. No physical evidence has been found to suggest she was killed during the terrorist attacks. [source: "Sneha Philip disappeared night before 9/11 amid allegations," *Daily Mail* (UK), Sep 11, 2019]

Priesner. Dr. Ernest Priesner (1934-1994?) was an Austrian biologist who was a leader in the field of insect pheromone (secreted chemical from an animal) research. He developed a set of 21 pheromones which attracts almost all type of insects and winged animals. In July 1994, he disappeared after leaving his house to collect insect traps in the Bavarian Alps. Various search operations failed to find any trace of him.

Rusanov. Vladimir Rusanov (1875-1913?) was a Russian geologist. In 1912, he was appointed to command a government expedition to Svalbard (Spitsbergen), an archipelago in the Arctic Ocean, to investigate the coal potential. The captain of the ship was Alexander Kuchin, Roald Amundsen's South Pole navigator. After 3

months, several expedition members returned to Russia. Without any approval, Rusanov and 11 others made an attempt to reach the Pacific Ocean via the Northern Sea Route. The last to be heard of the expedition was on September 27, 1912 when Rusanov telegraphed that he intended to go east across the Kara Sea to the Pacific Ocean. The ship and Rusanov's expedition were never heard of again. In 1937, relics of his expedition were found on Popova-Chukchina Island and the Mona Islands in the Kara Sea.

Weisfeiler. Dr. Boris Weisfeiler (1941-disappeared in 1985) was a Soviet-born mathematician and professor at Penn State University. In December 1984, he flew to Chile to hike alone in the Chilean Andes. The Chilean government said that he drowned in January 1985, his body never recovered. His backpack was found next to a river. Other sources say he was kidnapped, questioned, tortured by the Chilean military and executed. According to U.S. State Department reports, witnesses claim they saw Boris in the Colonia Dignidad German Colony in Chile several years after his disappearance. [source: "Chilean Court Upholds as Lawful the Disappearance, Kidnapping and Murder 30 Years Ago of Boris Weisfeiller," *concernedscientists.org*, Nov 30, 2019]

Williams. Dr. John C. P. Williams (1922-?) is a New Zealand cardiologist. In 1961, he discovered what is now called Williams syndrome, a genetic disorder caused by the deletion of about 27 genes. Williams had a reputation of being odd and eccentric. In September 1979. He renewed his passport in Geneva, and then disappeared. The High Court of New Zealand declared him a missing person in 1978. There is some evidence that he might be alive as recently as 2000. An author reported that Williams contacted him indirectly in January 2000.

Oddities of Science

Oddities of the Nobel Prize

Nobel-Prize Winning Scientists

Arrhenius. Svante Arrhenius (1859-1927) was a Swedish physicist and chemist. He was one of the founders of the science of physical chemistry. In 1896, he was researching climate change. He warned of the risks of the growing CO_2 emissions by man. In 1900, he was involved in setting up the Nobel Institutes and the Nobel Prizes. From 1900 to 1903, he was nominated for a Nobel Prize in Physics and in Chemistry 34 times. In 1903, he received the Nobel Prize for Chemistry, becoming the first Swedish Nobel laureate. He was the first to explain the fact that solid crystalline salts disassociated into paired charged particles when dissolved. In 1905, he became director of the Nobel Institute, where he remained until his death. He was a member of the Nobel Committee on Physics and on Chemistry.  He used his positions to arrange prizes for his friends (van't Hoff, Ostwald, and Richards) and to attempt to deny Nobel Prizes to his enemies (Ehrlich, Nernst, and Mendeleev). [source: "Svante Arrhenius, the Man Who Foresaw Climate Change," *OpenMind*, Feb 19, 2019]

Banting. Sir Frederick Grant Banting (1891-1941) was a Canadian physician. In 1923, Banting and John Macleod shared the Nobel Prize in medicine for their discovery of insulin. At age 32, he is the youngest Nobel laureate in the area of Physiology/Medicine. In February 1941, he died of wounds and exposure following a crash of a Lockheed Super Electra. After departing from Gander, Newfoundland, for a trip to London, both engines failed on the aircraft. Banting survived the initial impact, but died of his injuries the next day after a rescue team failed to reach him on time. [source: "Death of a Hero: Sir Frederick Banting's Plane Crash," *Hidden Newfoundland*, Dec 12, 2017]

Barany. Dr. Robert Barany (1876-1936) was an Austro-Hungarian otologist (ear doctor). He was nominated for the Nobel Prize in

Medicine 7 times between 1910 and 1914. He received the 1914 Nobel Prize in Physiology or Medicine for his work on the physiology and pathology of the vestibular apparatus (part of the middle ear).

During World War I, he served in the Austro-Hungarian Army as a civilian surgeon. He was later captured by the Imperial Russian Army. When the Nobel Prize was awarded in 1914, Barany was in a Russian POW camp. In 1916, he was released from the POW camp following diplomatic negotiations with Russia by the Prince of Sweden and the Red Cross. He was then able to attend the Nobel Prize awards ceremony in 1916, where he was awarded his prize. [source: Bracha & Tan, "Robert Barany: The Nobel Prize-winning prisoner of war," *Singapore Medical Journal*, January 2015]

Bardeen. Dr. John Bardeen (1908-1991) was an American physicist and electrical engineer. He was nominated for the Novel Prize in Physics 16 times between 1953 and 1966. He is the only person to be awarded the Nobel Prize in Physics twice, in 1956 for the invention of the transistor, and again in 1972 for his work in superconductivity, which led to the invention of MRI and CAT scans. He became the first person to win two Nobel Prizes in the same field. [source: "PBS American Portrait: John Bardeen," *pbs.org*, 1999]

Behring. Dr. Emil A. von Behring (1854-1917) won the 1901 Nobel Prize in Physiology or Medicine, the first one awarded, for his discovery of a diphtheria antitoxin in 1893. At the time, over 60,000 children died from diphtheria in Germany. His Nobel Prize medal is on display at the International Red Cross Museum in Geneva. Von Behring is believed to have cheated Paul Ehrlich (1854-1915) and Shibasaburo Kitasato (1853-1931) out of sharing the Nobel Prize. Ehrlich cooperated with Behring to create an antiserum for treating diphtheria and tetanus.

Behring schemed against Ehrlich at the Prussian Ministry of Culture. By 1900, Ehrlich refused to work with Behring. However, in 1908, Ehrlich did receive the Nobel Prize in Physiology or Medicine for his contributions to immunology. Kitasato was actually nominated for the Nobel Prize as he worked with Behring in Berlin and was the one that announced the discovery of the diphtheria antitoxin serum. [source: "Emil Adolf Behring," *Famous Scientists*]

Born. Max Born (1882-1970) was a German-Jewish physicist who was instrumental in the development of quantum mechanics. For years, he was ignored by the Nobel Committee. He was nominated for the Nobel Prize in Physics 34 times between 1930 and 1954. Finally, in 1954, he won the Nobel Prize in Physics for his fundamental research in quantum physics and his probabilistic treatment of the wave function. Born was a lover of music. His granddaughter was Olivia Newton-John (1948-). [source: "Max Born, the Quantum Physicist who Believed that 'God Plays Dice," bbvaopenmind.com, Jan 8, 2018]

Butenandt. Adolf E. Butenandt (1903-1995) was a German biochemist who was awarded the Nobel Prize in Chemistry in 1939, along with Leopold Ruzicka (1887-1976) for his work on sex hormones. In 1935, he synthesized the male hormone testosterone. He was nominated 18 times for a Nobel Prize in Chemistry or Medicine between 1934 and 1939. He had to reject the award in accordance with the Nazi government policy at the time. He accepted the award after World War II in 1949. [source: Adolf Butenandt Is Dead at 91; Won Nobel for Hormone Work," *New York Times*, Jan 19, 1995, Section B, p. 11]

Chalfie, Shimomura, and Tsien. In 2008, the Nobel Prize in Chemistry was awarded to Martin Chalfie, Osamu Shimomura (1928-2018), and Roger Tsien for their work on green fluorescent protein, or GFP. GFP is used to make biosensors. However, American molecular biologist Dr. Douglas Prasher (1951-) was the first to clone the GFP gene and should have been awarded the Nobel Prize. Dr. Prasher's accomplishments were not recognized, and by 2008, he was working as a courtesy shuttle bus driver for Toyota in Huntsville, Alabama for $8.50 an hour. Chalfie, Shimomura, and Tsien

all acknowledged Prasher's work and all agreed that he should have won the Nobel Prize in Chemistry. But the Nobel Prize states that no more than three people can share a Nobel Prize. Shimomura was lucky to survive World War II. A few days before his 17th birthday, he lived 15 miles from Nagasaki when the second atomic bomb was dropped and exploded over Nagasaki. The explosion temporarily blinded him and was later drenched in the radioactive "black rain" bomb fallout.

Curie, Marie. Marie Sklodowska Curie (1867-1934) was a Polish and naturalized French physicist and chemist. She coined the word "radioactivity." Marie Curie was a French citizen, but proud of her  Polish roots. When she and her husband discovered a new element in 1898, she named it "polonium" in honor of Poland. She was the first woman to win a Nobel Prize. She is still the only person to win Nobel Prizes in two different sciences: in physics in 1903 (shared with her husband, Pierre, and Henri Becquerel) and chemistry in 1911. The Curie family has won 5 Nobel Prizes. The 1903 Nobel Committee only intended to give the prize to Pierre Curie and Becquerel. However, a committee member and advocated for women's rights, alerted Pierre Curie to the situation. After his complaint, Marie's name was added to the nomination. Pierre Curie disliked public ceremonies, so the Curies declined to go to Stockholm to receive their prize. However, as Nobel laureates were required to deliver a lecture, the reluctantly went to Stockholm in 1905. During World War I, she developed mobile radiography units to provide X-ray services to field hospitals. In 1934, she died of leukemia, almost certainly caused by her extensive exposure to radiation. [source: Inan, "15 Mathematical Curiosities to Celebrated Marie Curie's 150th Birthday," *Scientific American*, Nov 6, 2017]

Domagk. Gerhard Domagk (1895-1964) was a German pathologist and bacteriologist who was awarded the Nobel Prize in Medicine in 1939 for his 1935 discovery of the first drug effective against bacterial infections. He was forced to decline the award due to Nazi policy. He is credited with the discovery of the first commercially available an-

Oddities of Science

tibiotic marketed under the brand name Prontosil. He treated his own daughter with the antibiotic, saving her the amputation of an arm. After World War II, he accept in Nobel Prize in 1947, but not the monetary portion of the prize due to the time that had elapsed. He was nominated for a Nobel Prize in Medicine 9 times between 1938 and 1953. [source: "Gerhard Domagk," *Science History Institute*, Dec 4, 2017]

Einstein. Albert Einstein (1879-1955) won the Nobel Prize in Physics in 1901 for his discovery of the law of the photoelectric effect. There was no explanation as to why the photoelectric effect worked at the time. It wasn't until 1924 that there was a universal acceptance of photons. Einstein did not get any money from his Nobel Prize.  In 1919, he signed his divorce papers to get out of his troubled marriage with his first wife, Mileva Maric. He agreed that if he won the Nobel Prize, the money would go to her. Maric eventually invested the Nobel Prize money in three apartment buildings in Zurich and remained well-off the rest of her life. Einstein was nominated for the Nobel Prize in Physics 62 times between 1910 and 1922. [source: Laskow, "It Wasn't Relativity That Won Einstein His Nobel Prize," *The Atlantic*, Sep 19, 2014]

Fibiger. Johannes Fibiger (1867-1928) was a Danish physician. He won the 1926 Nobel Prize in Physiology or Medicine for his discovery that a roundworm could cause cancer. His experimental results were later proven to be a case of mistaken conclusion. Erling Norrby, who served on the Nobel Committee declared that Fibinger's Nobel prize was the biggest blunder of the Nobel awards. After Fibiger's death, it was discovered that the roundworm did not cause cancer. He was nominated for the Nobel Prize in Medicine 18 times between 1920 and 1928. A month after winning the Nobel Prize, he died of a heart attack. [source: "An analysis of a wrong Nobel Prize – Johannes Fibiger," *Advances Cancer Research*, 2004]

Forssmann. Dr. Werner Forssmann (1904-1979) was a German physician that shared the 1956 Nobel Prize in Medicine for developing a procedure that allowed cardia catherization. In 1929, heart surgery was still in its infancy and doctors struggled to treat cardiac patients invasively. Forssmann suspected that he could reach the heart by snaking a hollow tube through the veins, but doctors told him that the procedure would be fatal. An operating-room nurse, Gerda Ditzen, agreed to sneak him some sterile supplies as long as he promised to perform the procedure on her instead of himself. Forssmann agreed, but tricked her. He restrained her to the operating table and pretended to locally anesthetize and cut her arm whilst doing it on himself. He anesthetized his own lower arm and inserted a urinary catheter into his vein. Not knowing of the catheter might pierce a vein, he put his life at risk and blindly guided the catheter into the right ventricle cavity of his own heart. Triumphant and still alive, he hobbled own to the X-ray lab to prove that the procedure worked. He was later forced to leave the hospital for self-experimentation. He was fired several times as a cardiologist and he eventually became a urologist. From 1932 to 1945, he was a member of the Nazi Party. He was captured by the Americans, and put into a U.S. prisoner of war camp. Upon release in 1945, he worked as a lumberjack for 5 years before returning to medicine. [source: Sudlow, "Heart pioneers and the curious case of Werner Forssmann," *British Heart Foundation*, Dec 11, 2018]

Haber. Fritz Haber (1868-1934) was a German chemist who received the Nobel Prize in Chemistry in 1918 for his invention of a method to synthesize ammonia from nitrogen gas and hydrogen gas. His process enabled the production of fertilizer in quantities that revolutionized agriculture worldwide. He actually received the award in 1919. During the selection process in 1918, the Nobel Committee for Chemistry decided that none of the year's nominations met the criteria as outlined in the will of Alfred Nobel. According to the statutes, the Nobel Prize can be reserved until the following year, and this statute was then applied. Haber, therefore, received his prize for 1918 one year later in 1919. Haber was nominated for the Nobel Prize in Chemistry 10 times between 1912 and 1919. Haber is also considered the "father of chemical warfare" for his years of pioneering work developing and weaponizing chlorine and other poisonous gases during World War I. Haber married Clara Immerwahr, the first woman to earn a PhD in chemistry. In 1915, following an argument with Haber, Clara committed suicide in their garden by shooting herself in the heart. [source: King, "Fritz Haber's Experiments in Life and Death," *Smithsonian Magazine*, June 6, 2012]

Kilby. Jack Kilby (1923-2005) was an American electrical engineer who helped create the first integrated circuit in 1959 while at Texas Instruments. Kilby conceived and built the first electronic circuit in which all of the device's active and passive components were fabricated in a single semiconductor substrate (made of germanium) half the size of a paper clip. He won the 2000 Nobel Prize in Physics for his work on the integrated circuit. He is also the co-inventor of the handheld or pocket calculator and the thermal printer. He created the first computer incorporating integrated circuits. [source:  Kabel, "Kilby surprised to win Nobel Prize as IC inventor," *EE Times*, Oct 10, 2000]

Kuhn. Richard Kuhn (1900-1967) was an Austrian-German biochemist who was awarded the Nobel Prize in Chemistry in 1938 for his work on vitamins and organic pigments called carotenoids. From 1900 to 1918, Kuhn was a schoolmate of Wolfgang Pauli, who was awarded the Nobel Prize in Physics for 1945. He had to reject the prize as Adolf Hitler had forbidden German citizens to accept it. He received the award after World War II in 1948. He had been nominated 8 times for a Nobel Prize in Chemistry from 1932 to 1939. [source: "Richard Kuhn, Biochemist, Dies; Was Denied Nobel Prize by Nazis," *New York Times*, Aug 2, 1967, p. 37]

Laue. (1879-1960) was a German physicist who won the Nobel Prize in Physics in 1914 for his discovery of the diffraction of X-rays by crystals. He was a strong objector to Nazism and was instrumental in re-establishing and organizing German science after World War II. In 1935, Max von Laue and James Franck sent their 23 karat gold medals to Niels Bohr's Institute of Theoretical Physics in Copenhagen for protection. When Nazi Germany invaded Denmark in World War II, the Hungarian chemist George de Hevesy (1885-1966) dissolved the gold medals of Laue and James Franck (1882-1964) in aqua regia (nitric acid and hydrochloric acid) to prevent the Nazis from discovering them. Hevesy placed the resulting solution on a shelf in his lab at the Niels Bohr Institute. After the war, he returned to find the solution still on the shelf. He was able to precipitate the gold out of the acid. The Nobel Society then re-cast the Nobel Prize gold medals, using the original gold. These recast medals were presented to von Laue and Franck in 1952. [source: Gignac, "The Invisible Prize," *American Institute of Physics*]

Lenard. Philipp von Lenard (1862-1947) was a Hungarian-born German physicist. In 1905, he won the Nobel Prize in Physics for his work on cathode rays and the discovery of many of their properties. He discovered that the energy (speed) of the electrons ejected from a cathode depends only on the wavelength, and not the intensity of the incident light. He was a German nationalist and anti-Semite. He was an active proponent of the Nazi ideology and was a Hitler supporter in the 1920s. He labeled Einstein's contribution to science as "Jewish physics." He despised "English physics," which he considered to have stolen its ideas from Germany. Under Hitler, he became Chief of Aryan Physics. He accused the Jewish press of promoting Einstein's dangerous work on relativity. Einstein responded, "When you are courting a nice girl, an hour seems like a second, but when you sit on a red-hot cinder, a second seems like an hour. That's relativity." He was nominated 15 times for a Nobel Prize in Physics from 1901 to 1925. [Gunderman, "When Science Gets Ugly – The Story of Philipp Lenard and Albert Einstein," *The Conversation*, June 16, 2015]

Marconi. Guglielmo Marconi (1874-1937) was an Italian inventor and electrical engineer. He was nominated for a Nobel Prize in Physics 15 times, from 1901 to 1933. He shared the 1909 Nobel Prize in Physics for his contributions to the development of wireless telegraphy. In his Nobel Prize acceptance speech, he freely admitted he didn't really understand how his invention worked. In 1888, Heinrich Hertz (1857-1894) demonstrated that one could produce and detect electromagnetic radiation. At the time, the radiation was commonly called "Hertzian" waves, and is now called radio waves. Marconi experimented and was able to transmit signals only up to one half mile. He got greater range by raising the height of the antenna and grounding his transmitter and receiver. He could now transmit up to 2 miles. By 1901, he could transmit messages about 2,200 miles away using a 500-foot kite-supported antenna. When Marconi won the Novel Prize, Nikola Tesla protested, saying that he had developed a wireless telegraph in 1893. In 1943, the U.S. Supreme Court invalidated four Marconi patents, citing Tesla's prior work. 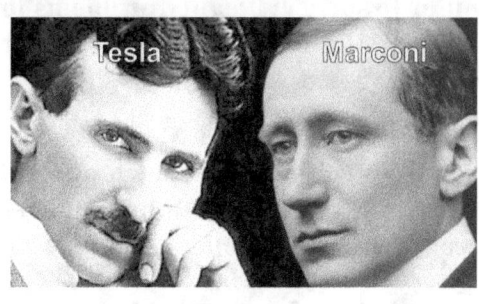 [source: Robinson, "Marconi forged today's interconnected world of communication," New Scientist, Aug 10, 2016]

Marshall. Barry James Marshall (1951-) is an Australian physician who won the 2005 Nobel Prize in Medicine for his discovery on a bacteria that caused gastritis and peptic ulcer disease. He was able to prove that the bacterium Helicobacter pylori played a major role in causing peptic ulcers, challenging decades of medical doctrine that ulcers were caused by stress, spicy foods, and too much acid. For years, he watched in horror as ulcer patients fell so ill that many had their stomach removed or bled until they died. At the time, ulcers afflicted 10% of all adults. He was told that bacteria couldn't survive in the human stomach. He was prohibited from experimenting on people, so he decided to experiment on himself. He took some H. pylori from the gut of an ailing patient, stirred it into a broth, and drank the bacteria. He expected to develop, perhaps a few years later, an ulcer. He was surprised when, only 3 days later, he developed gastritis (massive inflammation), the precursor to an ulcer. He was nauseous, started vomiting, his breath began to stink, and he felt sick and exhausted. He then biopsied his own gut, culturing H. pylori and proving that bacteria were the underlying cause of ulcers. He had a full recovery after antibiotic therapy. [source: "The Doctor Who Drank Infectious Broth, Gave Himself an Ulcer, and Solved a Medical Mystery," *Discover Magazine*, Apr 8, 2010]

Mayer. Dr. Maria Goeppert-Mayer (1906-1972) was a German-born American theoretical physicist. She was nominated 27 times for the Nobel Prize in Physics and in Chemistry from 1955 to 1963. In 1963, she won the Nobel Prize in Physics for proposing the nuclear shell model of the atomic nucleus, which she published in 1950. She was the second woman to win a Nobel Prize in Physics, after Marie Curie. She later programmed the Aberdeen Proving Ground's ENIAC computer to solve criticality problems for a liquid metal cooled reactor using the Monte Carlo method. [source: "Maria Goeppert-Mayer," Scientific Women, Apr 26, 2017]

Michelson. Albert Michelson (1852-1931) was the first American to win a Nobel Prize in science when he was awarded the 1907 Nobel Prize in Physics "for his optical precision instruments and the spectroscopic and metrological investigations carried with their aid."

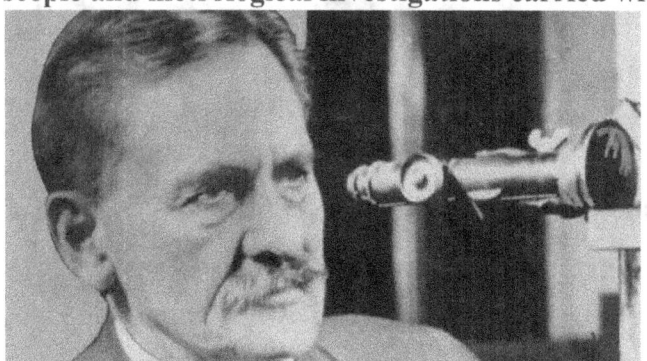

Michelson had devised his own device called an interferometer, which could measure the speed of light. The funds to build the instrument came from Alexander Graham Bell. In 1887, with the help of chemist Edward Morley (1838-1923) they had their results. There was no change in the speed of light from any direction at any time. This is not what the two scientists expected to find. Their experiment for the expected motion of the Earth relative to the ether (the hypothetical medium in which light was supposed to travel), resulted in a null result. One scientist called the Michelson-Morley interferometry experiment "the most famous negative results in the history of physics." [source: Livingston, *The Master of Light: A Biography of Albert A. Michelson*, 1979]

Moniz. Antonio Egas Moniz (1874-1955) was a Portuguese neurologist and the developer of the cerebral angiography, which provides images of blood vessels in and around the brain. In 1935, he developed the frontal lobotomy surgical procedure using an ice pick, for which he won a Nobel Prize in Medicine in 1949, the first Portuguese national to receive a Nobel Prize. His work would no longer be prizeworthy from today's perspective. Critics accused Moniz of understating complications, providing inadequate documentation, and not following up with patients. He was nominated for the Nobel Prize in Medicine 18 times from 1928 to 1950. From the 1930s to the 1980s,

Oddities of Science

50,000 lobotomies were performed in the United States. [source: Glorfeld, "The man who perfected the lobotomy," *Cosmos*, April 3, 2019]

Mullis. Kary Banks Mullis (1944-2019) shared the 1993 Nobel Prize in Chemistry with Michael Smith for his invention of the polymerase chain reaction (PCR) technique. It is used to make many copies of a specific DNA in a test tube rather than in an organism. It is a central technique in molecular biology which allows for the amplification of specified DNA sequences. Dr. Mullis wrote fiction and managed a bakery for two years before working at a biotechnology company as a DNA chemist. He invented a UV-sensitive ink and became skeptical of the existence of the ozone hole and climate change. He became skeptical that HIV caused AIDS. He founded a business to sell pieces of jewelry that had DNA from deceased famous people like Elvis Presley and Marilyn Monroe. He asserted his belief in astrology. [source: Grens, "Kary Mullis, Inventor of the PCT Technique, Dies," *The Scientist*, Aug 12, 2019]

Pauling. Linus Pauling (1901-1994) was an American chemist, biochemist, chemical engineer, and peace activist. He published over

1,200 papers and books. Pauling was the first to describe the alpha-helix structure of protein molecules. In 1953, Pauling rushed to publish is DNA theory of a triple helix, which turned out to be fatally flawed. Crick and Watson discovered the right theory, that DNA was a double helix. He was nominated 80 times in Chemistry, and in Medicine, and for the Peace Prize from 1940 to 1963. In 1954, he was awarded the Nobel Prize in Chemistry for his work in the chemical bond and the structure of molecules and crystals. In 1962, he was awarded the Nobel Peace Prize. He is one of 4 individuals to have won more than one Nobel Prize (the others are Marie Curie, John Bardeen, and Frederick Sanger). He is the only person to have been awarded two unshared Nobel Prizes. He and Marie Curie are the only two people to be awarded Nobel Prizes in different fields. Pauling is responsible for the widespread misbelief that high doses of vitamin C are effective against colds and other illnesses. No responsible medical or nutrition scientists share these views. Because of his anti-war demonstrations, the U.S. State Department temporarily withdrew his passport. Rumors circulated that he was a communist. When he won the 1962 Peace Prize for his antinuclear campaign, critics called him a naïve spokesman for the Communist Party. [source: Barankin, "The Impact of Linus Pauling on Modern Medicine and Society," *Bulletin of Anesthesia History*, October 1999]

Pavlov. Dr. Ivan Pavlov (1849-1936) was a Russian physiologist. He  was nominated for a Nobel Prize in Medicine 64 times, from 1901 to 1930. In 1904, he was awarded the Nobel Prize for Physiology or Medicine, becoming the first Russian Nobel laureate. Starting in 1901, Pavlov was nominated for a Nobel Prize four years in a row. He did not win the prize until 1904 because his previous nominations were not specific enough to any discovery, but base on a variety of laboratory findings. In 1904, it was specified in recognition for his work in the physiology of digestion. He measured the saliva of dogs and analyzed the saliva and what response it had to food under different conditions. By  the way, Pavlov never trained a dog to salivate to the sound of a bell. His Nobel Prize money was confiscated as property of the Russian state. From 1917 to 1920, after the Bolshevik Revolution, Pavlov struggled to feed his own family and to keep from freezing. At least a third of Pavlov's scientific colleagues died in those post-revolution years. Some scientists starved to death in apartments just above and below Pavlov's in Petrograd (Leningrad/St Petersburg). [source: Specter, "Drool: Ivan Pavlov's real quest," *The New Yorker*, Nov 17, 2014]

Oddities of Science

Planck. Max Planck (1858-1947) was a German physicist whose discovery of energy quanta won him the Nobel Prize in Physics in 1918. He was nominated for the Nobel Prize in Physics 74 times, from 1907 to 1919. Tragically, Planck would live to see the death of his wife and all 4 of their children. In 1909, his first wife died from tuberculosis. During World War I, his son Erwin (1893-1945) was taken prisoner by the French in 1914. His oldest son, Karl (1888-1916) was killed in action at Verdun on the Western Front in France. His daughter, Grete (1889-1917), died in 1917 during childbirth. His daughter Emma (1889-1919) also died in childbirth two years later. In January 1945, his son, Erwin, was sentenced to death by the Nazis for his role in the participation of the failed attempt to assassinate Hitler. He was executed for treason on January 23, 1945. Early in 1944, his home in Berlin was flattened in an allied air raid. All of Planck's personal papers and scientific records were destroyed. After his death, The Max Planck Society was formed, running over 80 scientific institutions. Research workers of the Max Planck Society have earned 18 Nobel prizes. [source: "Max Planck," *famousscientists.org*]

Roentgen. On November 8, 1895, Wilhelm Conrad Roentgen (1845-1923) was working with a cathode ray tube. Despite the tube being covered with heavy black cardboard, Roentgen noticed a fluorescent screen three feet away was glowing. The rays were somehow illuminating the screen. When he placed his hand in front of the tube, 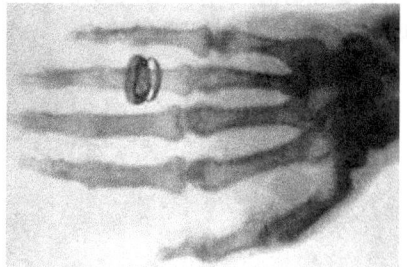 he noticed that he could see his bones in the image that was projected on a screen. He replaced the screen with a photographic plate to capture the images, creating the first X-ray photograph. He named it X-radiation to signify an unknown type of radiation. Roentgen then had a picture of his wife's hand on a photographic plate formed due to X-rays. It was the first photograph of a human body part using X-rays, which showed her bones and her wedding ring. When she saw the picture, she said, "I have seen death." He was nominated for the Nobel Prize in Physics and in Medicine 21 times, between 1901 and 1922. In 1901, he won the first Nobel Prize in Physics for his discovery of the "Roentgen ray." He donated the monetary award from his Nobel Prize to his university. [source: Nicholls, "Wilhelm C. Roentgen discoverer of X-rays," *European Heart Journal*, May 21, 2019]

Shockley. William Shockley (1910-1989) was an American physicist. In 1956, he shared the Nobel Prize in Physics for his work on semiconductors and the discovery of the transistor effect. Late in his life, he became interested in questions of race, human intelligence, and eugenics. He argued that people with low IQ (below 100), mainly African Americans, should be paid to undergo voluntary sterilization. Shockley was convinced that race-based IQ differences existed and spent most of his career after the 1960s promoting his racist theories and a high IQ sperm bank. [source: Moffitt, "How a racist genius created Silicon Valley by being a terrible boss," SFGate, Aug 22, 2018]

Thomson. Sir Joseph John Thomson (1856-1940) was a British physicist. In 1897, he discovered the electron, the first subatomic particle discovered. At the time, he called them 'corpuscles.' He was nominated for the Nobel Prize in Physics 24 times, from 1902 to 1912. He was awarded the 1906 Nobel Prize in Physics for his work on the conduction of electricity in gases. Eight of his students won Nobel Prizes (Charles Barkla, Niels Bohr, Max Born, William Bragg, Owen Richardson, Charles Wilson, Francis William Aston, and Ernest Rutherford). In addition, his son, George, won the 1937 Nobel Prize in physics for his work on the electron. [source: "J. J. Thomson," *biography.com*, June 25, 2019]

Zewail. Ahmed Hassan Zewail (1946-2016) was an Egyptian-American chemist who won the 1999 Nobel Prize in Chemistry for his work on femtochemistry (chemistry that studies very fast chemical reactions, about 10^{-15} seconds, or one femtosecond). He is the first Egyptian and the first Arab to receive a science Nobel Prize. [source: "Ahmed Zewail, Nobel laureate who sparked a 'revolution in chemistry, is dead at 70," *Washington Post*, Aug 3, 2016]

Scientists Who Should Have Won a Nobel Prize

Avery. Dr. Oswald Avery (1877-1955) was a Canadian-American physician and medical researcher. He was one of the first molecular biologists and a pioneer in immunochemistry. In 1943, he isolated DNA as the material of which genes and chromosomes are made. Several Nobel laureates said that Avery was the most deserving scientist not to receive the Nobel Prize for his work. He was nominated for the

Oddities of Science

award throughout the 1930s, 1940s, and 1950s. ["Oswald T. Avery, the unsung hero of genetic science," *The Guardian*, June 3, 2013]

Beijerinck. Dr. Martinus Beijerinck (1851-1931) was a Dutch microbiologist and botanist. He is considered one of the founders of virology and environmental (soil) microbiology. He was the first person to use the term "virus." He was the first to isolate a wide range of microorganisms. His idea that a pathogen can be a soluble molecule that proliferates when it is part of the protoplasm of a living cell was revolutionary and new. This new concept laid the foundation of virus research. In spite of his pioneering and number of contributions to science in general, he was never awarded the Nobel Prize. His advisor, Jacobus van't Hoff (1852-1911), won the first Nobel Prize in Chemistry in 1901. [source: van Kammen, "Beijerinck's contribution to the virus concept," *Archives of Virology*, 1999]

Bier. . August Bier (1861-1949) was a German surgeon who could have won a Nobel prize, but Adolf Hitler forbade German from accepting it. Instead, he was one of only 9 people ever to receive the German National Prize for Art and Science, an award created in 1938 by Adolf Hitler in replacement for the Nobel Prize. Bier was the first to perform spinal anesthesia and intravenous regional anesthesia. In 1898, he theorized that he could inject cocaine into the space around the spinal cord and numb his patients for surgery without putting them to sleep. At first, Bier asked his assistant, August Hildebrandt, to inject liquid cocaine into his spine, but the assistant fumbled the procedure. So, he numbed his assistant's legs to see whether or not he could feel the pain. Bier then proceeded to stick needles in his legs, hit his shin with an iron hammer, put cigar burns on his legs, and other pain-inducing actions. Bier's procedure worked. Bier recommended the procedure for surgeries of legs, but gave it up due to the toxicity of cocaine. Bier went on to treat many important people, such as Kaiser Wilhelm II, family members of the Czar of Russia, and Vladimir Lenin. [source: Erjavic, "August Karl Gustav Bier | The Embryo Project," *embryo.asu.edu*, Nov 15, 2017]

Burnell. Jocelyn Bell Burnell (1943-) co-discovered the first radio pulsars in 1967. In 1974, Antony Hewish was a co-winner of the 1974 Nobel Prize in Physics for his role in the discovery of pulsars, but it was Jocelyn, Hewish's graduate student, who as the first to notice the stellar radio source that was later recognized as a pulsar. Many people argued that she should have won a Nobel Prize, but Bell Burnell herself said that it would demean Nobel Prizes if they were awarded to research students. However, at least 8 research students have won Nobel Prizes. In 2004, Bell Burnell was recognized with a $3 million Special Breakthrough Prize in Fundamental Physics. She donated it all to assist female, minority, and refugee students in becoming physics researchers. [source: Drake, "Meet Jocelyn Bell Burnell, the Pioneer Who Found the Most Useful Stars in the Universe," *National Geographic*, Sep 6, 2018]

Cushing. Harvey Cushing (1869-1939) was an American neurosurgeon and pathologist. He was a pioneer in brain surgery and considered the father of neurosurgery. He developed a method of operating on the brain with local anesthesia. From 1917 to 1939, he 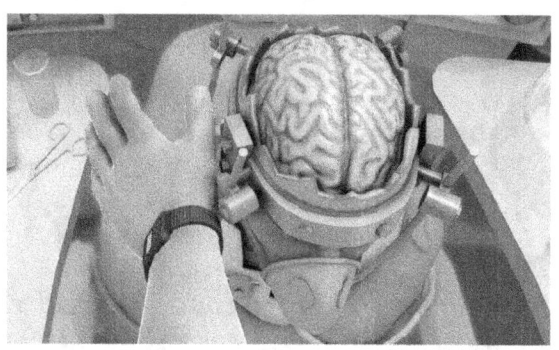 received 38 nominations for the Nobel Prize in Medicine, but never won. [source: Hansson, ""Highly qualified loser"? Harvey Cushing and the Nobel Prize, *ncbi.nih.gov*, Jan 2, 2015]

Damadian. Dr. Raymond Damadian (1936-) is an American physician and inventor of the first Magnetic Resonance (MR) Scanning Machine in 1974. In 2003, the Nobel Prize in Physiology or Medicine was awarded to Paul Lauterbur (1929-2007) and Sir Peter Mansfield (1933-2017) for their discoveries related to MRI. Damadian was not included in the prize. There was a suggestion that Damadian might have been denied a Nobel prize because of his creationist views. Today, there are over 23,000 MRI machines around the world that were used in over 60 million examinations. [source: "Inventor of the MRI denied the 2003 Nobel Prize in Medicine," *fonar.com*]

Eyring. Henry Eyring (1901-1981) was a Mexico-born, American theoretical chemist who authored over 600 scientific articles. He developed the Absolute Rate Theory of Transition state theory of chemical reactions, one of the most important developments in 20th-century chemistry. He allegedly failed to receive a Nobel Prize because of his membership in The Church of Jesus Christ of Latter-day Saints, or the Mormons. His father, who practiced polygamy in Mexico, was married to two daughters of Miles Romney, the great-grandfather of Mitt Romney. As partial compensation for not winning the Nobel Prize, the Nobel academy awarded him the Berzelius Medal in 1977. [source: H. J. Eyring, *Mormon Scientist*, 2007]

Franklin. Rosalind Franklin (1920-1958) was an English chemist and took the first X-ray diffraction images of DNA to prove it was a double helix. The images were used by James Watson (1928-) and Francis Crick (1916-2004) as the two co-authored the paper proposing the double helix structure of the DNA molecule. Both were awarded the Nobel Prize in 1962, but Franklin had died a few years earlier from ovarian cancer. Nobel Prizes cannot be awarded posthumously. It is highly plausible that, were she alive, she would have shared the Nobel Prize. [source: "Sexism in science: did Watson and Crick really steal Rosalind Franklin's data?"*The Guardian* (UK), June 23, 2015]

Grassi. Giovanni Grassi (1854-1925) was an Italian physician and zoologist. His works in malaria remain a lasting controversy in the history of Nobel Prizes, because he should have shared the 1902 Nobel Prize in Physiology or Medicine. Earlier, British army surgeon Ronald Ross (1857-1932) discovered the transmission of malarial parasite in birds and was awarded the 1902 Nobel Prize. However, in

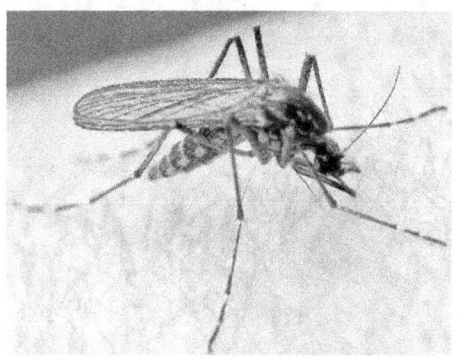

1898, it was Grassi who demonstrated the complete route of transmission of human plasmodium (living structure of cytoplasm), and correctly identified the types of malarial parasite as well as the mosquito. Grassi showed that the malarial parasites were carried only by the female mosquito. Grassi was fortunate to have access to sites where malaria was present near Rome and Sicily. When the 1902 Nobel Prize in Physiology or Medicine was considered, the Nobel Committee initially intended the prize be shared between Ross and Grassi. But Ross protested and accused Grassi of deliberate fraud without any proof. By today's standards, Grassi would have won a Nobel Prize. Grassi was later made a senator in Italy by King Victor Emmanuel III. [source: Cox, "History of the discovery of the malaria parasites and their vectors," *Parasites & Vectors*, Feb 1, 2010]

Hubble. Edwin Hubble (1889-1953) is regarded as one of the most important astronomers of all time. He discovered that many objects thought to be clouds of dust and gas were actually galaxies beyond our own Milky Way Galaxy. He proved that the universe was expanding. He showed that redshift increases with distance. The Hubble Space Telescope is named after him. Despite all of his accomplishments in astronomy he never received a Nobel Prize. At the time, the Nobel Prize in Physics did not recognize work done in astronomy. Hubble spent much of his later part of his career attempting to have astronomy considered an area of physics, instead of being its own science. He 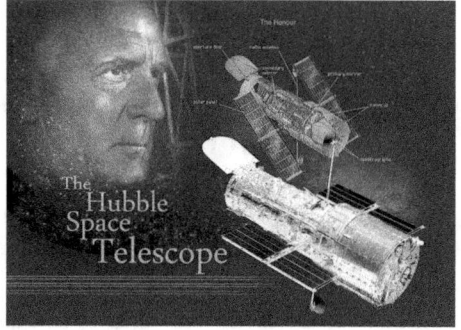 did this so that astronomers could be recognized for their valuable contributions to astrophysics. The campaign was unsuccessful in Hubble's lifetime, but shortly after his death, the Nobel Prize Committee decided that astronomical work would be eligible for the physics prize. It wasn't until 1974 that the Nobel Prize in Physics was awarded to an astronomer. That year, astronomers Martin Ryle and Antony Hewish won the Nobel Prize in Physics for their pioneering research in radio astrophysics and the discovery of pulsars. In 2011, the Hubble Telescope helped Adam Riess win the Nobel Prize in Physics in 2011 for his supernovae research. [source: Soares, "Hubble's Nobel Prize," *Journal of the Royal Astronomical Society of Canada*, March 20, 2001]

Ingold. Sir Christopher Kelk Ingold (1893-1970) was a British chemist. He is regarded as one of the chief pioneers of physical organic chemistry. He received 68 nominations for the Nobel Prize in Chemistry between 1940 and 1965, but never won. [source: Ridd, "Christopher Ingold: the Missing Nobel Prize, *pubs.acs,org*, 2017]

Leriche. Rene Leriche (1879-1955) was a French surgeon and physiologist. He was a specialist in pain and vascular surgery. He received 79 nominations for the Nobel Prize in Medicine from 1930 to 1953, but never won.

Oddities of Science

Lewis. Gilbert Lewis (1875-1946) was an American physical chemist. He discovered the covalent bond and his concept of electron pairs. He contributed to the valence bond theory which shaped modern theories of chemical bonding. In 1926, he coined the term "photon" for the smallest unit of radiant energy. It appeared in a letter to *Nature* magazine. He did major research on relativity and quantum physics. He was the author of 168 scientific publications. Despite all this, he never won a Nobel Prize. Yet, from 1922 to 1946, he was nominated a record 41 times for the Nobel Prize in Chemistry. There is evidence that one of the Swedish members of the Nobel Chemistry Committee did not like Lewis and wrote negative reports about him. Lewis mentored and influenced a number of Nobel Laureates, including Harold Urey, William Giauque, Glenn T. Seaborg, Willard Libby, and Melvin Calvin. [source: "Gilbert Newton Lewis," *sciencehistory.org*, Dec 11, 2017]

Meitner. Lise Meitner (1878-1968) was an Austrian-Swedish physicist who contributed to the discovery of nuclear fission of uranium. Between 1924 and 1965, she was nominated 19 times for the Nobel Prize in Chemistry, and 29 times for the Nobel Prize in Physics. She was the first woman to become a full professor of physics in Germany. She lost these positions when the Nazis came to power because she was Jewish. In 1938, she fled to Sweden. In 1939, with a letter to the editor of *Nature*, she described the process of splitting uranium and even named it fission. [source: Jorgensen, "Lise Meitner – the forgotten woman of nuclear physics who deserved a Nobel Prize," *The Conversation*, Feb 7, 2019]

Mendeleev. Dmitri Mendeleev (1834-1907) was a Russian chemist. He is best remembered for formulating the Periodic Law and creating a version of the periodic table of elements. He never won a Nobel Prize. In 1906, he came within one vote of winning a Nobel Prize, but died the next year. The Nobel chemistry panel supported Mendeleev's award for a Nobel prize, but the awards committee ruled that his discovery was not recent enough to qualify him for consideration. It is speculated that one dissenter on the Nobel Committee (Swedish physical chemist Svante Arrhenius) disagreed with Mendeleev's work. [source: Sutton, "The father of the periodic table," *Chemistry World*, Jan 2, 2019]

Oppenheimer. Julius Robert Oppenheimer (1904-1967) was an American theoretical physicist and is sometimes called the "father of the atomic bomb." He was nominated for the Nobel Prize in Physics three times, in 1945, 1951, and 1967, but never won. Oppenheimer might have won a Nobel Prized for his work on gravitational collapse, neutron stars, and black holes, but died at the age of 62, just when his work was getting recognized. [source: *"J. Robert Oppenheimer,"* Biography, Feb 14, 2020

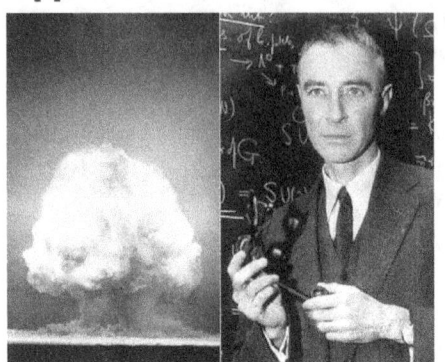

Poincare. Henri Poincare (1854-1912) was a French theoretical physicist, mathematician and engineer. In 1905, he first proposed gravitational waves. He received 51 nominations for a Nobel Prize in Physics from 1904 to 1912, but never won. Swedish mathematician Gosta Mittag-Leffler (1846-1927) campaigned hard to obtain Poincare's nomination for a Nobel Prize in Physics, but never succeeded. [source: Nastasi, "A Nobel Prize for Poincare?" *Lettera Matematica*, 2013]

Ramon. Gaston Ramon (1886-1963) was a French microbiologist and veterinarian who developed a vaccine against diphtheria and tetanus. At the time, diphtheria was a leading cause of death. His techniques are still used in vaccines manufactured today. He was nominated 155 time for the Nobel Prize in Physiology or Medicine, but never won a Nobel Prize. [source: Butler, "Close but no Nobel; the scientists who never won," *Nature*, Oct 11, 2016]

Roux. Pierre Paul Emile Roux (1853-1933) was a French physician and researcher who discovered the diphtheria bacterial toxin. He received 115 nominations for the Nobel Prize between 1901 and 1932, but never won. Roux was a co-founder of the Pasteur Institute. He also co-developed the rabies vaccine with Louis Pasteur.

Rubin. Dr. Vera Rubin (1928-2016) was an American astronomer who pioneered work on galaxy rotation rates. She was the first to provide evidence for the existence of dark matter. She was the first woman allowed to observe at Palomar Observatory. She never won the Nobel Prize in Physics, but several physicists and science writers have argued

Oddities of Science

that this was an oversight. [source: Childers, "Vera Rubin: The Astronomer Who Brought Dark Matter to Light," *space.com*, June 11, 2019]

Salk. Dr. Jonas Edward Salk (1914-1995) was an American medical researcher and virologist. He discovered and developed one of the first successful polio vaccines that was affecting over 58,000 cases a year (and over 3,000 deaths). Salk tested his vaccine on himself to be sure it was safe. Polio was considered one of the most frightening public health problems in the world. In the early 1950s, apart from the atomic bomb, America's greatest fear was polio, just like the coronavirus today. By April 2017, polio was eradicated from the planet. Salk never patented the vaccine and he never won a Nobel Prize. Although Salk never received a Nobel Prize for his work, the Salk Institute for Biological Studies has trained 5 Nobel Laureates. [source: Racaniello, "Polio and Nobel Prizes," *Virology Blog*, Sep 7, 2007]

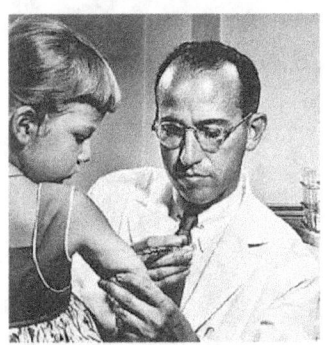

Sauerbruch. Ferdinand Sauerbruch (1875-1951) was a German surgeon who developed a pressure chamber for operating on the open thorax. It also allowed heart and lung operations to take place at greatly reduced risk. During World War I, he developed several new types of limb prostheses, which for the first time enabled simple movements with the remaining muscle of the patient. During World War II, he supported experiments of prisoners in concentration camps. He approved experiments with mustard gas on prisoners. In 1942, he became Surgeon General of the German army. In 1945, he was charged by the Allies for having contributed to the Nazi dictatorship, but not convicted for lack of evidence. Between 1914 and 1951, he received 51 nominations for the Nobel Prize in Medicine, but never won.

Schatz. Dr. Albert Israel Schatz (1920-2005) was an American microbiologist who should have won the Nobel Prize in Medicine. In 1952, Selman Waksman (1886-1973) won the Nobel Prize in Medicine for the discovery of the antibiotic streptomycin. The Nobel Prize committee ignored Waksman's graduate student, Schatz, who actually found the chemical after 3 and a half months of searching and isolating two distinct microorganisms excreting a substance that he called streptomycin.

Bill Wall

Oddities of Science

Oddities of Criminal Scientists

Bolber. Dr. Morris "The Rabbi" Bolber (1886- 1954) was a doctor for the "Philadelphia Poison Ring." . He was a serial killer that murdered over 50 people. In 1939 he was arrested and sentenced to life in prison. He was a part of a gang of 15 that could be hired for murder. The gang murdered at least 114 people, mainly through poison such as arsenic. The leaders of the gang were eventually caught and executed. [source: Snyder, "Bewitched: Witchcraft Life Insurance and the Business of Murder," *vtuhr.org,* 2017]

Crippen. Hawley Harvey Crippen (1862-1910) was an American ear and eye specialist. After a party at their home in England, his wife, Cora, disappeared. Dr. Crippen claimed that she had returned to the United States. He later said that she died and had been cremated. After being  questioned by the police, he went to Antwerp and boarded a Canadian Pacific ocean liner. The police at Scotland Yard inspected his home and found his wife's body buried under the brick floor of the basement. She was poisoned and dismembered. Inspectors sent a wireless message to the ocean liner notifying the captain of a suspected murderer aboard. Crippen was arrested as the ship entered the St. Lawrence River. He was returned to England, went to trial, and was hanged for the murder of his wife. He was the first criminal ever to be arrested with the aid of wireless radio. [source: "Was Dr. Crippen innocent of his wife's murder?" *BBC*, July 29, 2019]

Davalloo. Sheila Davalloo (1969-) was an Iranian-American biochemist and pharmaceutical researcher who stabbed to death Anna Raymundo, her former lover's girlfriend in 2002. She worked at Purdue Pharma in Stanford, Connecticut. She was sentenced to 50 years in prison. [source: Hayden, "Sheila Davalloo Maintains Innocence In Romantic Rival's Slaying: 'I'm Going to Uncover The Truth;," *Oxygen*, Dec 1, 2019]

Donovan. John J. Donovan , Sr. (1942-) is a former management professor at MIT. He authored multiple computer science textbooks. In 2005, he was accused of falsely claiming that 2 men attacked him and that his son orchestrated the shooting. John Sr. decided to shoot himself in the stomach as part of a bizarre attempt to convince authorities that his son had laundered $180 million and hired the Russian mob to put a hit on his old man. Donavan tampered with a security camera, shot his own car full of holes, simulated a break-in of his house, and falsified a police report in order to implicate his son. He was subject to a lifetime Stay Away Order and must never be within 100 yards of his children. In 2017, he was indicted on 7 counts of forgery and a single count of attempted larceny for faking the signature of his late son to the codicil of his will. [source: Macquarrie, "DA Says Former Professor Staged Attack On Himself,: *The Tech* (MIT), May 5, 2006]

Fuchs. Dr. Klaus Fuchs (1911-1988) was a German physicist who became an atomic spy. He supplied information from the American, British, and Canadian Manhattan Project to the Soviet Union during and shortly after World War II. Fuchs joined the Communist Party and later fled Germany to Britain in the early 1930s during the rise of the Nazis. He earned his PhD in physics from the University of Bristol. From late 1941 until 1946, he passed atomic secrets to the Soviets as the operative "Rest." In January 1950, he confessed he was a spy and was sent to prison. He was released in 1959 and immigrated to East Germany. [source: Holmes, "Spies Who Spilled Atomic Bomb Secrets," *Smithsonian Magazine*, April 19, 2009]

Gluzman. Rita Gluzman (1948-) was a cancer researcher and co-owner of a computer company in New Jersey, just like her husband, Yakov. When Yakov threatened to divorce Rita for another woman, Rita hired her cousin, Vladimir Zelenin, to kill Yakov. In 1996, Yakov's body was stabbed, beaten with an axe, and then chopped into more than 65 pieces with a hacksaw and scalpel. Police caught Zelenin dumping garbage bags in the Passaic River that contained the chopped up parts of Yakov Gluzman. Zelenin told police that Rita made him kill Yakov so that Rita could retain control of their computer company, ECI Technologies. Rita was given a life sentence. Gluzman became the first woman to be convicted under the U.S. Violence Against Women Act. [source: Berger, "Woman Sentenced to Life for Ax Killing of Husband, *New York Times*, May 1, 1997]

Huang. Dr. Gurang Matthew Huang (1963-2002) was a molecular biologist who had run a prestigious genome center in China. He

Oddities of Science

worked as a director of molecular biology. In 2002, he gunned down 46-year-old geneticist Dr. Tanya Holzmayer, the co-inventor of a patented technique to screen genes in a search for cures to AIDS and cancer. Her murder was witnessed by her teenage son in Mountain View, California. He later took his own life in Foster City, where he had been living. [source: "Firing may have spurred slaying-suicide / Mountain View victim lured out be pizza delivery," *SFGate*, March 1, 2002]

Kevorkian. Dr. Jack Kevorkian (1928-2011) was an American pathologist and euthanasia proponent. He claimed to have assisted at least 130 patients in their deaths. He was often called "Dr. Death." In 1998 he was arrested in Michigan and tried for his direct role in a case of voluntary euthanasia on a man who was suffering from Lou Gehrig's disease. He was convicted of second-degree murder and was given a 25-year prison sentence. He was released after 8 years on good behavior. [source: "Jack Kevorkian – Doctor," *biography.com*, July 26, 2019]

Kontz. Ann Miller Brier Kontz (1970-) was a chemist and research scientist at drug maker GlaxoSmithKline. In 2000, she poisoned her husband, Dr. Eric Miller, with arsenic. She had the help of her lover and co-worker, Derril Willard, who later committed suicide a month after the murder. Eric was a pediatric AIDS researcher at UNC, Chapel Hill, She pleaded guilty and was sentenced to 31 years in prison. [sources: "Miller Poisoning Case Dramatically Changed Live," *wral.com*, Nov 13, 2005 and Smith, "Gorgeous Chemist Poisons Husband While Having Affair With Co-Worker," *Oxygen*, Nov 7, 2018]

Lieber. Dr. Charles Lieber is chair of the Department of Chemistry and Chemical Biology at Harvard. In 2020, he was arrested, accused of lying about his ties to China. He was criminally charged with making false, fictitious and fraudulent statements to the U.S. Defense Department about his ties to a Chinese government program to recruit foreign scientists and researchers. One source says that he was arrested for concealing funding from a Chinese lab in Wuhan, China, supposedly connected to the origin of the new coronavirus. [source: Chappell, "Acclaimed Harvard Scientist Is Arrested, Accused of Lying About Ties to China, *NPR*, Jan 28, 2020]

Lookman. In 2019, Turab Lookman, a former Los Alamos National Laboratory scientist, pleaded guilty of lying about his contact with a

Chinese recruitment program. Lookman is the author of two scientific books and over 250 academic articles. [source: Wyland, "Ex-LANL scientist pleads guilty to lying to government," *Santa Fe New Mexican*, Jan 24, 2020]

Rai. Chiman Rai (1941-) was a professor of Mathematics at Alcorn State University in Mississippi. In April 2000, he hired a hitman to murder his son's wife because she was black and not Indian. She was found stabbed and strangled to death in her Union City apartment. In 2015, he was convicted of murder, burglary, and related offenses in connection with the 2000 death of his daughter-in-law, Michelle "Sparkle" Reid Rai, age 22. Rai paid a hit man $10,000 to have her killed. [source: Corson, "How the AJC covered the Sparklke Rai case from 'ATL Homicide,'" *Atlantic Journal-Constitution,* Aug 6, 2018]

Roth. Dr. J. Reece Roth (1938-) is a former electrical engineering professor at the University of Tennessee. He had been the director of the Plasma Sciences Laboratory. In 2006, he was found by a federal jury to have violated the International Traffic in Arms Regulation (ITAR) by allowing foreign national students to work on research contracted by the U.S. Air Force. He was charged with conspiracy, wire fraud, and exporting defense articles and services without a license. [source: Rohrlich, "Air Force Scientist Spilled No Secrets. He Still Went to Prison," *Daily Beast*, May 27, 2018]

Sik. In 2019, Dr. Bulent Sik, a food engineer who worked with Turkey's Ministry of Health, was sentenced to 15 months for releasing findings of a study linking environmental factors and cancer cases. His report revealed the cancer risks posed by toxic pollution in western Turkey. [source: "Turkey doctor gets 15 months for revealing pollution cancer risk," *New Straits Times*, March 15, 2020]

Snook. James Snook (1879-1930) was a veterinarian. He was head of the Department of Veterinary Medicine at Ohio State University. He invented a surgical instrument for the use in spaying animals. In 1920, he was a member of the U.S. Olympic Pistol Team that won a Gold Medal at the 1920 Olympics in Antwerp Belgium. In 1929, he was having an affair with Theora Hix, a medical student, when she threatened to kill Snook's wife and daughter. Snook then smashed her head with a ball-peen hammer, then cut her carotid artery for her to die. Snook was later arrested, sentenced to die, and was executed in the electric chair in Ohio. [source: "James Howard Snook," *Murderpedia.org*]

Oddities of Science

Other Books by Bill Wall

01. 300 King's Gambit Miniatures	-1982	ISBN 0-931462-17-7	80 pages
02. 500 Sicilian Miniatures	-1983	ISBN 0-931462-24-X	107 pages
03. 500 French Miniatures	-1984	ISBN 0-931462-31-2	126 pages
04. 500 Queen's Gambit Miniatures	-1985	ISBN 0-931462-38-X	127 pages
05. Larsen's Opening (1.b3)	-1986	ISBN 0-931462-55-X	57 pages
06. Owen's Defense (1.e4 b6)	-1986	ISBN 0-931462-50-9	65 pages
07. 500 King's Gambit Miniatures	-1986	ISBN 0-931462-57-6	101 pages
08. 500 Ruy Lopez Miniatures	(1986, 1997)	ISBN 0-931462-56-8	119 pages
09. 500 Italian Miniatures	-1987	ISBN 0-931462-65-7	101 pages
10. 500 Sicilian Miniatures II	(1987, 1997)	ISBN 0-931462-73-8	107 pages
11. Grob's Attack (1.g4)	-1988	ISBN 0-931462-86-X	84 pages
12. 500 Queen's Gambit Miniatures, II	-1988	ISBN 0-931462-87-8	108 pages
13. The Orangutan (1.b4)	-1989	ISBN 0-931462-92-4	90 pages
14. 500 Indian Miniatures	-1990	ISBN 0-931462-99-1	98 pages
15. 500 English Miniatures	-1990	ISBN 0-945070-04-5	102 pages
16. 1990 World Chess Championship	-1991	ISBN 0-945470-08-8	90 pages
17. 500 Caro Kann Miniatures	-1991	ISBN 0-945470-15-0	102 pages
18. Smith-Morra Accepted	-1992	ISBN 0-945470-22-3	118 pages
19. Smith-Morra Declined	-1993	ISBN 0-945470-25-8	120 pages
20. 500 Pirc Miniatures	-1993	ISBN 0-945470-38-X	86 pages
21. 500 Alekhine Miniatures	-1994	ISBN 0-945470-38-X	104 pages
22. Dunst Opening (1.Nc3)	-1995	ISBN 0-945470-48-7	106 pages
23. 500 French Miniatures II	-1995	ISBN 0-945470-54-1	117 pages
24. 500 King's Gambit Miniatures II	-1996	ISBN 0-945470-61-4	111 pages
25. 500 Scotch Miniatures	-1997	ISBN 0-945470-89-4	106 pages
26. 700 Opening Traps	-1998	ISBN 0-945470-71-1	107 pages
27. 500 Blackmar-Diemer Miniatures	-1999	ISBN 0-945470-80-0	94 pages
28. 500 Center Counter Miniatures	-2001	ISBN 0-945470-85-1	116 pages
29. Off The Wall Chess Trivia e-book - Pickard & Son	-2001		287 pages
30. Winning with the Krazy Kat and Old Hippo -Lulu	-2008	(ID:3292224)	124 pages
31. 700 Opening Traps	2019	ISBN 9781708915124	151 pages
32. Mate in Three	2019	ISBN 9781675985496	160 pages
33. Chess Opening Blunders	2020	ISBN 9781660738885	188 pages
34. Oddities in Chess	2020	ISBN 9798608854613	180 pages
35. 600 Sicilian Miniatures	2020	ISBN 9798623526014	154 pages

Bill Wall's Oddities Series

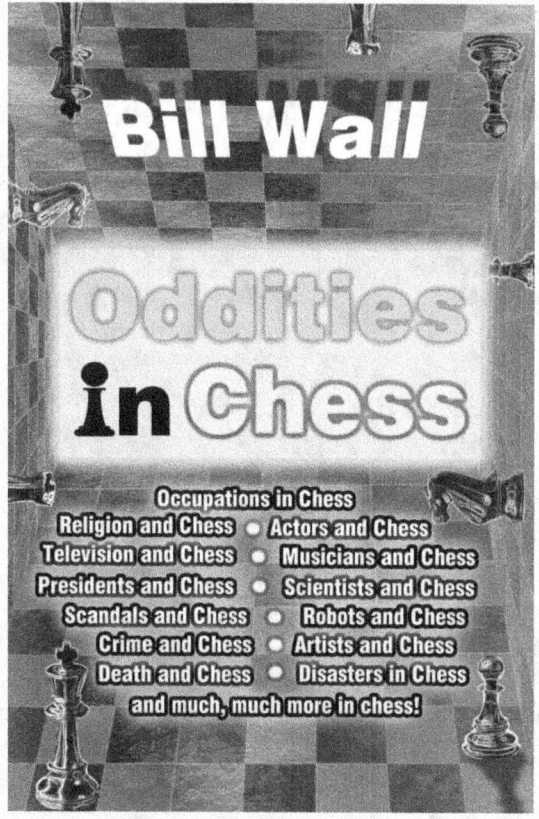

See Bill Wall's recent books on Amazon

www.amazon.com/author/billwall